ENVIRONMENTAL SCIENCE, ENGINEERING AND TECHNOLOGY

CADMIUM TOXICITY IN CROPS:
A REVIEW

ENVIRONMENTAL SCIENCE, ENGINEERING AND TECHNOLOGY

Additional books in this series can be found on Nova's website under the Series tab.

Additional E-books in this series can be found on Nova's website under the E-books tab.

Environmental Science, Engineering and Technology

Cadmium Toxicity in Crops: A Review

Conceição Santos
Marta Monteiro
and
Maria Celeste Dias

Nova Science Publishers, Inc.
New York

Copyright © 2010 by Nova Science Publishers, Inc.

All rights reserved. No part of this book may be reproduced, stored in a retrieval system or transmitted in any form or by any means: electronic, electrostatic, magnetic, tape, mechanical photocopying, recording or otherwise without the written permission of the Publisher.

For permission to use material from this book please contact us:
Telephone 631-231-7269; Fax 631-231-8175
Web Site: http://www.novapublishers.com

NOTICE TO THE READER

The Publisher has taken reasonable care in the preparation of this book, but makes no expressed or implied warranty of any kind and assumes no responsibility for any errors or omissions. No liability is assumed for incidental or consequential damages in connection with or arising out of information contained in this book. The Publisher shall not be liable for any special, consequential, or exemplary damages resulting, in whole or in part, from the readers' use of, or reliance upon, this material.

Independent verification should be sought for any data, advice or recommendations contained in this book. In addition, no responsibility is assumed by the publisher for any injury and/or damage to persons or property arising from any methods, products, instructions, ideas or otherwise contained in this publication.

This publication is designed to provide accurate and authoritative information with regard to the subject matter covered herein. It is sold with the clear understanding that the Publisher is not engaged in rendering legal or any other professional services. If legal or any other expert assistance is required, the services of a competent person should be sought. FROM A DECLARATION OF PARTICIPANTS JOINTLY ADOPTED BY A COMMITTEE OF THE AMERICAN BAR ASSOCIATION AND A COMMITTEE OF PUBLISHERS.

LIBRARY OF CONGRESS CATALOGING-IN-PUBLICATION DATA
Santos, Conceigco.
  Cadmium toxicity in crops : a review / authors: Conceigco Santos, Marta Monteiro and Maria Celeste Dias.
    p. cm.
  Includes index.
  ISBN 978-1-61728-169-3 (softcover)
  1. Plants--Effect of cadmium on. 2. Soils--Cadmium content. I. Monteiro, Marta. II. Dias, Maria Celeste. III. Title.
  QK753.C16S26 2010
  632'.19--dc22

Published by Nova Science Publishers, Inc. ✝ New York

CONTENTS

Preface		vii
Introduction		1
Chapter 1	Cadmium in Soils and Uptake by Plants	3
Chapter 2	Cadmium Phytotoxicity	7
Chapter 3	Plant Metal Accumulation and Hyperaccumulation	29
Chapter 4	Trophic Transfer, Bioaccumulation and Biomagnification of Cadmium	33
Chapter 5	Conclusions and Future Prospects	37
References		41
Index		65

PREFACE

Soil pollution by metals is a major environmental challenge. Metals are transmitted from plants to humans *via* the food chain. In particular, cadmium (Cd), a non essential metal, is listed as a "priority substance" by EC (2455/2001/EC, 2001) and by Environmental Protection Agency (EPA, USA), and is released to environment mainly due by anthropogenic activities.

Remediation strategies claim a better knowledge on Cd phytotoxicity and detoxification mechanisms. Cadmium toxicity involves, for example, growth reduction, wilting, photosynthesis and consequently carbon metabolism decreases. At the cell level, recent data support that Cd may have cytotoxic effects, in particular leading to oxidative stress, with impairment of antioxidative enzymes. Cd induced-genotoxicity is not so well studied, but some studies in crops suggest that Cd may induce DNA degradation (clastogenicity), may decrease DNA content and may have mitogenic effects. Moreover, molecular techniques (e.g. microsatellites, AFLP, RAPDs) showed to be powerful tools to assay Cd genotoxicity, with several studies supporting Cd induced point mutations.

This book revises some of the most recent advances concerning Cd induced effects at cytotoxicological and genotoxicological levels in several crops (e.g. *Lactuca* sp. and *Brassica* sp.) and in other model species as *Thlaspi* sp. and *Arabidopsis* sp.. Particular emphasis is also given to the whole plant performance, concerning this metal toxicity, including gas-exchange and chlorophyll a fluorescence parameters, carbohydrates, Calvin cycle enzymes and pigments content. The book also revises some important aspects concerning Cd accumulation and distribution in plants and their relevance to the trophic transfer of Cd.

Finally, the potential use of new techniques as flow cytometry (quantitative cytometry) and molecular markers are critically explored in these kinds of toxicological assays, and their potential use to develop new endpoints in ecotoxicology is also revised.

INTRODUCTION

Metals are among the most persistent environmental pollutants, and although some are essential plant micronutrients (e.g. copper or iron) others such as cadmium (Cd) have no known biological functions, except for the recently described Cd carbonic anhydrase in marine diatoms (Lane and Morel 2000; Lane et al. 2005). Cadmium is of worldwide concern because it is a toxic trace pollutant for humans, animals and plants. It enters the environment mainly from industrial processes and phosphate fertilizers, and is then transferred to the food chain. Due to its known toxicity, Cd is listed as "priority substance" by EC (2455/2001/EC 2001) and by Environmental Protection Agency (EPAUSA).

Cadmium occurs naturally in soils (e.g. in complexes), but it has been estimated that the anthropogenic emissions of Cd are in the range of 30 000 t per year (Shutzendubel and Polle 2002), mostly due to mining activities, burning of fossil fuels (ATSDR 1999), or metallurgical industry (Robards and Worsfold 1991). In agricultural soils, treated sewage sludge ("biosolids") and phosphate fertilizers (Speir et al. 2003; McLaughlin et al. 2006; Singh and Agrawal 2007) are also important sources of Cd contamination. Unlike other metals (as mercury (Hg) or lead (Pb)), Cd has relative mobility in soils and does not bind strongly to organic matter (Nelson and Campbell 1991). Besides its long environmental persistence, Cd has an extremely long biological half-life that essentially accounts for its bioaccumulation in individuals (e.g. John and Leventhal 1996). Cadmium has been largely studied and it was demonstrated to be cytotoxic, mutagenic and/or carcinogenic (e.g. Filipic et al. 2006) in animal cells. Despite less studied in plants, available data also support Cd has a cytotoxic and genotoxic environmental pollutant. We will explore in this review some of the recent research on Cd phytotoxicity (mostly directed to cytotoxicity and genotoxicity), and critically evaluate the use of new techniques used to Cd cytotoxicity and/or genotoxicity. Furthermore, the implications of metal accumulation in plants to the trophic transfer of this metal are highlighted.

Chapter 1

CADMIUM IN SOILS AND UPTAKE BY PLANTS

Cadmium has relative mobility in the plant-soil system. Despite the fact that Cd concentrations in non-polluted soils are variable and dependent on sources of minerals and organic material, it was estimated that on average they contain 0.04 to 0.32 µM of Cd (Wagner 1993). Soil solutions with Cd concentrations from 0.32 to 1 µM are considered moderately polluted (Sanitá di Toppi and Gabbrielli 1999).

Cadmium toxicity has been disclosed as early as 1955 in Japan as Itai–itai disease, and for the first time, Cd pollution was shown to have severe consequences on human health (Bertin and Averbeck 2006). Cadmium levels up to 800 mg/kg have been reported for soils in polluted areas (IARC, 1993). In a Korean mining area, Jung and Thornton (1996) found Cd concentrations up to 40 mg/kg in surface soils. Also, Cd concentrations of 0.5 mg/kg or more have been found in rice grown in Cd-polluted areas of Japan (Nogawa et al. 1989) and China (Cai et al. 1990). Furthermore, in a recent field study in Europe performed by Peris et al. (2007) the Cd content in edible parts of vegetables such as lettuce were found to be above the maximum levels established by the Commission Regulation no. 466/2001 for horticultural crops (466/2001/EC, 2001).

Plants may accumulate Cd mostly by root uptake, but also by foliar uptake. The degree to which plants are able to uptake Cd, is conditioned by its concentration in the soil, and its bioavailability, modulated by the presence of organic matter, pH, redox potential, temperature and concentrations of other elements (Benavides et al. 2005). Cadmium uptake by plants grown in contaminated soils has been studied, mostly in sludge-amended soils (e.g.

McLaughlin et al. 2006, Singh and Agrawal 2007) and in soils treated with Cd-enriched phosphate fertilizers (e.g. Huang et al. 2003). For example, Cd bioavailability increases with decreasing pH in soil, but Cd uptake may be reduced at low pH because of competition with H^+ ions at root uptake sites (Greger 1999). Acid rain may acidify soil and surface waters increasing the geochemical mobility of Cd (Campbell 2006). Contrarily, by binding Cd ions, organic matter may decrease Cd uptake (He and Singh 1993, Prasad 1995). Chloride levels may also affect Cd availability as soil sodium chloride has an antagonistic effect on metal toxicity (Bhartia and Singh 1994). Finally, the rhizosphere may have a pH different from the soil directly around it, and therefore can significantly affect the uptake of most plant nutrients.

Cadmium co-existence in soil solution with other essential and non-essential metals leads to various synergic and antagonistic interactions conditioning plant uptake and tissue content (Palóve-Balang et al. 2006). Cadmium ions may compete with macro and micronutrients and, such as calcium (Ca) and zinc (Zn), for the same transmembrane carriers (Sanitá di Toppi and Gabbrielli 1999), putatively leading to nutrient deficiencies (Krupa et al. 2002).

Once inside the root, and similarly to other mineral ions, Cd moves across the root through the symplast and/or the apoplast (Sanitá di Toppi and Gabbrielli 1999) until it reaches the endoderm. The endodermal cell layer acts then as a barrier for apoplastic diffusion into the vascular system, forcing most ions to enter the symplastic pathway prior entering the xylem (Jacoby and Moran 2002, McLaughlin 2002). Subsequently, three processes govern the movement of metals from outer root regions to xylem: sequestration on metals inside the cells, symplast transport into the stele and release into the xylem (Benavides et al. 2005). Cell membranes play, therefore, a key role in metal (and particularly Cd) homeostasis, preventing or reducing entry into the cell. On other hand, the apoplast continuum of the root epidermis and cortex is readily permeable to solutes.

The mechanisms regulating metal transport across the plasma membrane to the stele are still not fully understood (McLaughlin 2002). For cationic metals, such as Cd, the main route for uptake across the plasma membrane is the large negative electrochemical potential produced as a result of the membrane H^+ translocating adenosine triphosphatase (ATPases) (McLaughlin 2002). For example, in lettuce plants exposed to increasing Cd concentrations, it was shown that high amounts of Cd in roots were correlated with high contributions from H^+-ATPase in the active process of Cd uptake (Costa and Morel 1994). Other studies suggest that the main route for uptake of divalent metals is via ion channels, such as Cd^{2+} and Mg^{2+} channels (McLaughlin 2002).

Subsequent to metal uptake into the root symplasm, three processes govern the movement of metals from the root into the xylem: sequestration of metals inside root cells, symplastic transport into the stele and release into the xylem (Clemens et al. 2002). During their transport through the plant, metals become bound to cell walls, which can explain why normally Cd^{2+} ions are mainly retained in the roots, and only small amounts are translocated to the shoots (Cataldo et al. 1983, Greger 1999). But once loaded in the xylem sap, Cd is translocated to the aerial parts of plants through the transpiration stream, where they might be present as a divalent ion (Greger 1999) or complexed by several ligands, such as amino acids, organic acids and/or, perhaps, phytochelatins (PC) (e.g. Sanitá di Toppi and Gabbrielli 1999, Gong et al. 2003).

Chapter 2

CADMIUM PHYTOTOXICITY

Plants represent unique models for *in situ* monitoring of soil contamination and can potentially be used as biomonitors of environmental quality. Moreover, the contamination of crops with metals (such as Cd), by soil or irrigation with contaminated water may raise the problem of trophic transfer and ultimately human contamination.

In some highly metal polluted regions, Cd phytotoxicity may lead to a decrease in agricultural crop productivity (Vassilev and Yordanov 1997). Different studies involving different plant species, Cd compounds and concentrations, demonstrated that Cd is strongly phytotoxic and causes in general growth inhibition and even plant death, although the mechanisms involved in its toxicity are still not completely understood. In fact, it is well demonstrated that Cd interferes at several physiological levels (e.g. it affects photosynthesis, respiration and nitrogen metabolism, it induces oxidative stress and genotoxicity) culminating in poor growth and low biomass production (Sanitá di Toppi and Gabbrielli 1999, Fodor 2002), and eventual senescence and death. Understanding Cd uptake/allocation and plant cytological and physiological responses is therefore a requisite for the long-term safety and conservation of agricultural resources and ecosystems.

Apart from the recently discovered Cd-dependent enzyme in diatoms stated above, Cd is considered a non-essential and toxic element without any other known biological role (Palóve-Balang et al. 2006), that can affect plants on various organizational and functional levels. Plants respond to metal accumulation by expressing various manifestations of toxicity that can be detected and analyzed at various levels of organization ranging from gross morphology to cellular, biochemical or molecular levels, and thus can be useful to monitor as well as assess environmental metal pollution. Moreover, the sedentary nature of plants is a major advantage of plant-based assays for monitoring toxic chemicals in the environment. As for other stresses, the type and/or degree of response(s) to metal stress depend not only on stress intensity and duration but also on the tissue type and the age of the plant.

2.1. CADMIUM CYTOTOXICITY

Cadmium stress usually leads to a battery of unspecific symptoms that include chlorosis, necrotic lesions, wilting, reddish coloration and growth reduction (e.g. Sanitá di Toppi and Gabbrielli 1999, Azevedo et al. 2005a-c). Disturbances in plant water relations are also widely known as one of the first effects of Cd toxicity. It was even suggested that water stress caused by Cd may lead to a cascade of physiological and metabolic processes, including photosynthesis impairment (Barceló and Poschenrieder 1990).

Among the several metabolic pathways that have shown to be highly affected by Cd toxicity, carbon metabolism and antioxidant system are particularly sensitive and are of particular importance. The first one is involved in supplying energy to the plant and the cells have to deal with the multiple direct/indirect effects of Cd in the photosynthetic apparatus, and the second pathway is crucial as the Cd-exposed cells have to develop efficient strategies to overcome the increase of oxidative stress. Below, we'll explore in more detail these two metabolic aspects.

Photosynthesis and Carbon Metabolism

The photosynthetic apparatus is particularly susceptible to Cd toxicity and the reduction in photosynthetic rate is a common response in plants exposed to Cd (Mysliwa-Kurdziel et al. 2002, Burzynski and Klobu 2004). Inhibition of photosynthetic rate could be due to several structural and functional disorders in the different parts of this process. Photosynthesis can be inhibited at several levels: photosynthetic pigments synthesis, electron transport, enzymes of the Calvin cycle, structural disorganization of chloroplasts and uptake of nutrients (Larbi et al. 2002, Mysliwa-Kurdziel et al. 2002, Vassilev et al. 2002, Mallick and Mohn 2003, Burzynski and Klobu 2004, Balakhnina et al. 2005, Han et al. 2006, López-Millán et al. 2009).

Photosynthetic pigments have been shown as one of the main targets of the toxic Cd action. One of the most usual symptoms of Cd stress is chlorosis of the leaves due to an impairment of photosynthetic pigment biosynthetic pathways (Baryla et al. 2001, Mysliwa-Kurdziel 2002, Carrier et al. 2003), but also to a strong interaction between Cd and Fe that reduces uptake of Fe and causes Fe deficiency in leaves as well as to enzymatic degradation (Krupa et al. 2002, Vassilev et al. 2002). Cadmium can alter chlorophyll biosynthesis by inhibiting protochlorophyllide reductase. Moreover, Cd^{2+}, like other metals, can interfere with photosynthetic pigments through the substitution of the Mg^{2+} ion in the chlorophyll molecules by Cd^{2+} (Mysliwa-Kurdziel 2002). These substituted chlorophylls have much lower fluorescence quantum yields when compared to Mg-chlorophylls (Krupa et al. 2002).

Several authors have reported decreased levels of chlorophyll pigments in different plant species treated with Cd (e.g. Vassilev et al. 2002, Pietrini et al. 2003, López-Millán et al. 2009). A recent report of Fagioni and Zolla (2009) revealed that spinach plants exposed to Cd develops chlorosis and this is prevalently localized in the basal leaves. The reduced chlorophyll a synthesis and photosynthesis leading to senescence and cell death of the spinach plants. In *Phaseolus vulgaris*, Cd induced decreases in leaf chlorophyll content and other photosynthetic pigment. This effect was more marked in plants exposed to 10 µM Cd than to 100 µM (Han et al. 2006). A similar effect was observed in sugar beet, and attributed to a Cd-induced Fe deficiency at low Cd treatments (Larbi et al. 2002). Also in *Lycopersicon esculentum* (Delpérée and Lutts 2008), *Zea* mays (Ekmekci et al. 2008), *Brassica napus* (Ben et al. 2009) and *Lactuca sativa* (Monteiro et al. 2009b) Cd toxicity induced a decrease in photosynthetic pigments. Since the chlorophyll content may directly influence the functioning of the photosynthetic apparatus and thus affect overall plant metabolism, it is considered a key factor when assessing the impact of Cd stress (Fodor 2002).

The ratio Chl *a*/Chl *b* is another related endpoint relevant for Cd toxicity assessment. Although, there is no known direct influence of metal ions on the process of transformation of Chl *a* to Chl *b*, changes on Chl *a*/Chl *b* ratio are commonly reported on metal stressed plants (Mysliwa-Kurdziel and Strzalka 2002). Both increases and decreases in this ratio have been found in plants treated with Cd^{2+} (Mysliwa-Kurdziel and Strzalka 2002 and references therein). In *Brassica juncea*, the excess levels of Cd decreased the contents of chlorophylls but increased the ratio of chlorophyll a/b (Singh and Tewari 2003).

Several investigations on the effects of Cd toxicity on plant photochemical processes have been performed. It is generally accepted that water oxidizing system of PSII (photosystem II) is affected by Cd by replacing the Mn^{2+} ions, thereby inhibiting the reaction of PSII (Krantev et al. 2008). Cadmium exerts multiple effects on both donor and acceptor sites of PSII. On the donor site, the presence of Cd inhibits the oxygen evolving cycle and, consequently, oxygen evolution; on the acceptor site, it inhibits electron transfer from Q^-_A to Q^-_B (Ekmekci et al. 2008). The maximum photochemical efficiency of PSII was found to be reduced in different plant species exposed to Cd (Burzynski and Klobus 2004, Balakhnina et al. 2005, He et al. 2008, Bi et al. 2009, Monteiro et al. 2009b). PSII is extremely sensitive to Cd and its function is inhibited to a much greater extent than that of PSI (Mallick and Mohn 2003, Ekmekci et al. 2008). Paľove-Balang et al. (2006) reported that PSI activity is only slightly affected by Cd. However a recent study of Fagioni and Zolla (2009) in spinach plants reports a high sensitivity of PSI to Cd. According to Siedlecka and Baszynski (1993), the disturbed functional activity of PSI is due to Cd-induced Fe deficiency that limits the level of ferredoxin and $NADP^+$-oxydoreduction. A report of Siedlecka and Krupa (1996) on the interactions between Cd and Fe in bean plants also confirms this suggestion.

Cd reduces the photochemical quenching while the non-photochemical quenching increased (Krupa et al. 1993, Linger et al. 2005, Ekmekci et al. 2008). Di Cagno et al. (2001) also confirm this findings in Cd treated sunflower plants.

Limitations of photosynthesis in Cd treated plants are mainly due to non-mesophyll factors (Vassilev et al. 2002, Zhu et al. 2005, He et al. 2008) and inhibition of RuBisCO (ribulose 1,5-bisphosphate carboxylase/ oxygenase) activity is considered the main target of Cd toxicity (Siedlecka et al. 1997, Krantev et al. 2008). Cd^{2+} ions interfere with RuBisCO activation, lower its activity and damage its structure by substituting for Mg^{2+} ions, which are important cofactors of carboxylation reactions, and may also shift RuBisCO activity towards oxygenation reactions (Siedlecka et al. 1997, Pietrini et al. 2003). Cadmium toxicity causes an irreversible dissociation of the large and small subunits of RuBisCO, thus leading to total inhibition of the enzyme (Malik et al. 1992).

Photosynthesis impairment due to RuBisCO inhibition was described for Cd-treated bean plants (Siedlecka et al. 1997), barley (Vassilev et al. 1997), sunflower (Di Cagno et al. 2001) and *L. esculentum* (Chaffei et al. 2004). In maize Cd-treated plants the photosynthesis impairment was correlated with a drop in the activities of RuBisCO and also PEPCase (phosphoenol-pyruvate carboxylase) (Krantev et al. 2008). Moreover, in pigeon pea and wheat plants, Cd seems to inhibit the activity of all enzymes of the Calvin's cycle (Malik et al. 1992). The negative effects of Cd can also be observed in carbohydrates metabolism (Delpérée and Lutts 2008, Dunwei et al. 2009) due to the inhibition of several enzymes such as RuBisCO (Mobin and Khan 2007, Krantev et al. 2008), fructose-6-phosphate kinase (Malik et al. 1992), fructose-1,6-bisphosphatase (Sheoran et al. 1990), $NADP^+$-glyceraldehyde-3-phosphate dehydrogenase (Sheoran et al. 1990), PEPCase (Krantev et al. 2008), aldolase (Sheoran et al. 1990), and carbonic anhydrase (Mobin and Khan 2007).

Profound anatomical changes in leaves and structural disorganization of chloroplasts are also in the basis of the inhibition of photosynthesis (Mysliwa-Kurdziel et al. 2002). Cadmium induces a disorder in chloroplast ultrastructure (Bi et al. 2009), reduction on its size, decrease the number of grana and tylakoids (Pietrini et al. 2003, Vassilev et al. 2002) and change the shape and internal organization of the thylakoid membranes and stroma (Gratão et al. 2009). According to Skorzinska and Baszynski (1993) stroma thylakoids degradation is more severe than grana degradation. In leaves of *Pisum sativum*, Cd produced a significant inhibition of growth as well as a reduction in the transpiration and photosynthesis rate, chlorophyll content and an alteration in the nutrient status. The ultrastructural analysis of *P. sativum* grown with 50 µM $CdCl_2$, showed cell disturbances characterized by an increase of mesophyll cell size, and a reduction of intercellular spaces, as well as severe disturbances in chloroplast structure (Sandalio et al. 2001). Also on oilseed rape (*B. napus*), leaf chlorosis was attributable to a marked decrease in the chloroplast density caused by a reduction in the number of chloroplasts per cell and a change in cell size (Baryla et al. 2001).

Another unfavorable effects of toxic metals on plants are the inhibition of the normal uptake and utilization of mineral nutrients (Sandalio et al. 2001, Fodor 2002, Vassilev et al. 2002). One of the crucial factors of Cd^{2+} influence on plant metabolism and physiological processes is its relationship with other mineral nutrients. As mentioned above, Cd^{2+} transport across cell membranes is most likely facilitated by metal transporters that normally act to mobilize essential metals. Thus, by substituting for essential divalent cations, Cd^{2+} limits their uptake. Alternatively, Cd^{2+} may bind to specific groups of proteins and lipids or channel proteins of membranes, thereby inhibiting transport and disturbing the uptake of many macro and micronutrients. Furthermore, destruction of the cell membranes can also alter the ratio of essential elements and cause the decrease in their content, thereby inducing nutrient deficiencies

(Cseh 2002). One of the most important mechanisms for impairment of the uptake of nutrients by Cd is via the inhibition of Fe transport into the shoot, which has a pronounced effect on many aspects of the structure and function of the photosynthetic apparatus (Krupa et al. 2002). The induced iron shoot deficiency reduces the pool of Fe-containing electron carriers in the photosynthetic electron transport chain, causes disorganization of the chloroplast structure and even reduces RuBisCO content (Siedlecka and Krupa 1998). Cadmium is also known to cause other important disturbances in nutrient levels that can severely affect normal plant metabolism. Specifically it can decrease the levels of magnesium (Mg), potassium, phosphate, Ca and Zn, and increase manganese content (Krupa et al. 2002).

Plant growth inhibition is a classical parameter commonly used in the assessment of Cd toxicity to plants (Linger et al. 2005, Monteiro et al. 2009b) and a recommended endpoint in standard tests for toxicity assessment (OECD, 2006). Apart from being an important indicator of toxicity at an individual level, growth inhibition is a non-specific manifestation of alterations at a biochemical level that are produced as a more specific response of plants to the particular stress. Cadmium induces a decrease and/or inhibition of plant growth as a consequence of CO_2 assimilation decline and/or inhibition (Larbi et al. 2002, Krantev et al. 2008, López-Millán et al. 2009), and of translocation of photosynthetic products (Linger et al. 2005) and cell division inhibition (Dalla Vecchia et al. 2005, Linger et al. 2005). Excessive concentrations of Cd (100 and 1000 µM) affected photosynthesis and consequently plant growth in *Medicago sativa*, *Raphanus sativus*, and *L. sativa* (Benzarti et al. 2008). Similar results were obtained by Monteiro et al. (2009b) in *L. sativa* with 100 µM $Cd(NO_3)_2$.

Oxidative Stress

Exposure to abiotic and biotic stress results, in general, at some stage of exposure, in an increase in reactive oxygen species (ROS) (e.g. Schützendübel and Polle 2002). In particular, metals stimulate the formation of ROS, either by direct electron transfer involving metal cations, or as a consequence of metal-mediated inhibition of metabolic reactions. Reactive oxygen species are generated in plant cells during normal metabolic processes, such as respiration and photosynthesis (Mittler 2002). The ROS intermediates are partially reduced forms of O_2 and typically result from the excitation of O_2 to form a singlet ($^1O_2$), or from the transfer of one to three electrons to O_2 to form, respectively, O_2^-, H_2O_2 or HO^- (Mittler 2002). Oxidative stress in plants occurs when the delicate balance between antioxidants and ROS is tilted in favour of ROS (Razinger et al. 2009).

Some ROS may function as important signalling molecules (e.g., by altering gene expression and modulating the activity of specific defence proteins), but they can be extremely harmful at high concentrations (Apel and Hirt 2004). Reactive oxygen species may lead to the oxidation of proteins, lipids and nucleic acids, and often culminating with, e.g., membrane damage, mutagenesis and inactivation of enzymes, and thus affecting cell viability (Apel and Hirt 2004). As a consequence, tissues injured by oxidative stress generally contain increased concentrations of carbonylated proteins and malondialdehyde (MDA) (e.g. Azevedo et al. 2005c).

Schützendübel and Polle (2002) distinguished three different molecular mechanisms of metal toxicity: (a) ROS production by auto-oxidation and Fenton reaction; (b) blocking of essential functional groups in biomolecules, a reaction that has been mainly been reported for non-redox-reactive metals such as Cd; (c) displacement of essential metal ions from biomolecules. However, metals without redox capacity such as Cd do not seem to act directly on the production of ROS via Fenton and/or Haber Weiss reactions (Dietz et al. 1999). Instead, Cd can expedite ROS production indirectly by, for example, substituting an essential micronutrient, or binding unspecifically to thiol groups disturbing various metabolic processes (Razinger et al. 2009). For example, Cd may disrupt the photosynthetic electron chain, leading to ROS production of O_2^- and $^1O_2$ (Dietz et al. 1999).

The ROS formed in response to Cd may differ in time, nature and localisation. Tobacco cells exposed to $CdCl_2$ developed successive waves of ROS differing in nature and subcellular localization: the first one consisted in the transient NADPH oxidase-dependent accumulation of H_2O_2 followed by the accumulation of O_2^- in mitochondria. A third wave of ROS consisted in fatty acid hydroperoxide accumulation and was concomitant with cell death (Garnier et al. 2006).

One of the plant responses to ROS production is the increase in anti-oxidant enzyme activities providing protection from oxidative damage induced by several environmental stress (Apel and Hirt 2004). A variety of proteins function as scavengers of superoxide and H_2O_2. Among the major ROS-scavenging enzymes in plants are catalase (CAT), peroxidase (POX) and superoxide dismutase (SOD) (Mittler 2002). The superoxide released by processes such as oxidative phosphorylation is first converted to H_2O_2 and then reduced originating water, in a detoxification pathway that involves multiple enzymes, with SOD catalysing the first step (superoxide conversion to H_2O_2) and then CAT and various POX removing H_2O_2 from the cell. This type of response seems however to be species, time and stress-condition dependent. For example, CAT activity has been shown to be suppressed in diverse plant species exposed to Cd (Chaoui et al. 1997, Chaoui and El Ferjani 2005). In sensitive barley cultivars, activities of the enzymes involved in ROS detoxification were markedly enhanced by increasing Cd supply, with the exception of CAT, while in the tolerant ones, there was either only a slight increase or no change in the antioxidative enzymes activities, and the differential Cd tolerance was not related to uptake or accumulation of Cd (Tiryakioglu et al. 2006).

In the newly discovered hyperaccumulator *Sedum alfredii* Hance, POX played an important role during Cd hyperaccumulation (Zhang and Qiu 2007). In *Arabidopsis thaliana*, Cd elevated the activities of FeSOD and MnSOD, but CuZnSOD activity was diminished in comparison with control (Drazkiewicz et al. 2007). Also in alfalfa Cd exposure led only to a mild increase of POX (and other antioxidant molecules) (Ortega-Villasante et al. 2007), while in transgenic tall fescue plants, Cd led to an overexpression of the CuZnSOD and POX genes, which are utilized in scavenging ROS. For CAT, Cd induced a CATA3 transcript level, but this effect was reverted by ascorbate (Lee et al. 2007).

The anti-oxidative battery also includes non-protein scavengers, including ascorbate (ASC) and glutathione (GSH) (Mittler 2002) that play a pivotal role in the response of plant cells to abiotic stress (Horemans et al. 2007). In fact, these compounds act in a network in complement with anti-oxidative enzymes. Ascorbate peroxidase (APX) and glutathione reductase (GR), as well as GSH, are important components of the ascorbate–glutathione cycle responsible for the removal of H_2O_2 in different cellular compartments (Jiménez-Ambriz et al. 2007). Glutathione is also the substrate for the biosynthesis of PC, which are involved in metal detoxification (Zenk 1996).

Comparing the hyperaccumulator *Arabidopsis halleri* with the non accumulator *A. thaliana*, Chiang et al. (2006) found that, among others, APX and MDAR4 (in the ascorbate-glutathione pathway) were expressed at higher levels in *A. halleri*. These changes were supported by higher enzymatic activities APX in the hyperaccumulator species. Also, in *A. thaliana*, Cd decreased GSH/GSSG ratio ((Drazkiewicz et al. 2007). In *Pisum sativum*, Cd reduced GSH and ASC contents, and CAT, GR, POX and CuZnSOD activities, while MnSOD activity increased. Comparatively, CAT and CuZnSOD were down-regulated at transcriptional level, while MnSOD,

FeSOD and GR were up-regulated (Rodriguez-Serrano et al. 2006). The same authors also reported that Cd induced ethylene and ROS, while nitric oxid (NO) decreased, which correlated with senescence processes induced by Cd (Rodriguez-Serrano et al. 2006). In *Coffea arabica* L. suspension cultures, Cd induced activities of SOD, CAT and GR (ensuring the availability of reduced GSH) while POX activity seemed inhibited (Gomes-Junior et al. 2006). Comparing the *Arabidopsis* mutant rax1-1 with the mutant Cd hypersensitive 2 plants, Ball et al. (2004) showed a direct link between GSH biosynthesis and stress defence. Other enzymes, such as lipoxygenase, have their activities stimulated by Cd exposure (Tamas et al. 2009).

Cadmium produces alterations in the functionality of membranes, often related with induced changes in lipid composition (Ouariti et al. 1997), and by affecting the enzymatic activities associated with membranes, such as the H^+-ATPase (Fodor 2002). Tommasini et al. (1998) analysed an ATP-binding-cassette (ABC) transporter of *A. thaliana* with high sequence similarity to the human (MRP1) and yeast (YCF1) glutathione-conjugate transporters and in the ycf1 mutant, Cd resistance was partially restored and glutathione-conjugate transport activity was observed. In the same species, tolerance to Cd of transgenic plants is correlated with decreased accumulation of lipid peroxidation-derived reactive aldehydes compared to wild-type plant, and this is probably related with an increased activity of Ath-ALDH3 that limits aldehyde accumulation and oxidative stress (Sunkar et al. 2003). Moreover, Cd uptake is apparently a prerequisite for the change in dehydroascorbate transport activity, and this seems to be independent of the Cd-induced H_2O_2 production (Horemans et al. 2007). Transcript profiles of roots of *A. thaliana* and *T. caerulescens* plants exposed to Cd and Zn showed differences in gene expression between both species, mostly genes involved in lignin, GSH and sulphate metabolism (van Mortel et al. 2008).

Many plant metal transporters remain to be identified at the molecular level. Thomine et al. (2000) isolated AtNramp cDNAs from *Arabidopsis*, which expresses in both roots and aerial parts. AtNramp3 gene leads to slightly enhanced Cd resistance of root growth, and overexpression of AtNramp3 results in Cd hypersensitivity of *Arabidopsis* root growth (Thomine et al. 2000). Shigaki et al. (2005) identified in an *Arabidopsis* mutant a vacuolar transporter CAX1 with higher transport for Cd. Also, screening Cd-responsive genes in *Arabidopsis,* Susuki et al. (2002) identified a gene encoding a putative metal binding protein CdI19, which, upon introduction into yeast cells, conferred marked toleration of Cd exposure. The authors proposed that CdI19 plays an important role in the maintenance of metal homeostasis and/or in detoxification by endowing plasma membranes with the capacity to serve as an initial barrier against inflow of free metal ions into cells. Using a functional cloning strategy, Li et al. (2002) described for *A. thaliana* a detoxifying carrier AtDTX1 localised in the plasma membrane that mediates the efflux of plant derived or exogenous toxic compounds from the cytoplasm and that was capable of detoxifying Cd. Clemens et al. (1998) tested putative plant cation transporters for Cd uptake activity and suggested that LCT1 may represent a plant cDNA encoding a plant Ca cation uptake or an organellar Ca transport pathway in plants contributing to transport of Cd^{2+} across membranes.

The interaction of Cd with other nutrients in the antioxidative response is being largely demonstrated: a model was proposed by *Rodríguez-Serrano et al.* (2009), for the cellular response to long-term Cd exposure involving Ca, ROS and NO. Using pea (*P. sativum*), these authors demonstrated that Cd affected differentially the expression of SOD isozymes at both transcriptional and posttranscriptional levels, leading to SOD activity reduction. Calcium down regulated CuZnSOD, while NO synthase-dependent production was depressed by Cd exposure. Also Aravind et al. (2009) showed that Zn supplementation inhibited the oxidative products of proteins. Also Zn protection against Cd-induced ROS was confirmed by the reduced levels of oxidative products of proteins (Aravind et al. 2009). Moreover, the inhibition of ROS in Cd production in isolated plasma membranes was reversed by Ca and Mg (Heyno et al. 2008). In *B. napus*, selenium applied separately or in combination with Cd did not affect the activity of antioxidative enzymes in roots, but diminished it in the shoot (Filek et al. 2008). Finally, available data suggest that Cd, when not detoxified rapidly enough, may trigger, via the disturbance of the redox control of the cell, a sequence of reactions leading to growth inhibition, stimulation of secondary metabolism, lignification, and subsequent cell death (Schützendübel and Polle 2002).

The role of organelles in the metabolic pathways related with oxidative stress (e.g. fatty acid beta-oxidation, photorespiration, and metabolism of ROS and RNS) induced by Cd have been studied. It was described for *Arabidopsis* that ROS appeared first in the mitochondria and subsequently in the chloroplast (Bi et al. 2009). The same authors also described that previous treatment with ascorbic acid or CAT decreased the production of ROS and prevented inhibition of photosynthesis and organelle changes (Bi et al. 2009). Cadmium influenced differently ROS production in isolated plasma membranes and in mitochondria, in these last ones with increased levels of H_2O_2. *In vivo*, Cd stimulated ROS production in the mitochondrial electron transfer chain and inhibition of NADPH oxidase activity in the plasma membrane (Heyno et al. 2008). Rodríguez-Serrano *et al.* (2009) reported that treatment of plants with $CdCl_2$ significantly increased peroxisome speed, which was dependent on endogenous ROS and Ca, but was not related to actin cytoskeleton modifications. The same authors also proposed that this increase in peroxisomal motility may protect the cell against the Cd-imposed oxidative stress (*Rodríguez-Serrano et al.* 2009). Another example of Cd toxicity in cell components, is, besides the nucleus (and DNA content that will be explored below), the effect at the microtubular cytoskeleton, as microtubule alterations appeared after treatment with low Cd concentration, pointing to microtubules or microtubule-associated proteins, among the main targets of Cd (Fusconi et al. 2007).

The above described responses are merely a fraction of the multiple responses of plants to Cd stress, still yet to unfold. The dimension of the pathways putatively involved in response to Cd stress is evident in the still few works on genes responses. For example, in tolerant genotypes of backcrosses (BC1) between A. *halleri* ssp. *halleri* and the non-tolerant relative *A. lyrata* ssp. *petraea*, a hundred and thirty-four genes that expressed more than sensitive genotypes were identified, and most belonged to diverse functional classes, including ROS, detoxification, cellular repair, metal sequestration, water transport, signal transduction, transcription regulation, and protein degradation (Craciun et al. 2006). Chiang et al. (2006) used an *Arabidopsis* cDNA microarray to compare the gene expression of the Zn/Cd hyperaccumulator *A. halleri* and a non-hyperaccumulator, *A. thaliana* for better understanding the hyperaccumulating mechanism, and proposed genes that could be candidates for the future engineering of plants with large biomass for use in phytoremediation. Also, transcriptome analyses of the antioxidative enzymes in leaves of pea plants grown in the presence of Cd and treated with some modulators of the signal transduction cascade suggested the existence of cross-talk between these elements and ROS metabolism during Cd stress. At least Ca channels, phosphorylation/dephosphorylation processes, NO, cGMP, salicylic acid (SA) and H_2O_2 were involved in some steps between the Cd signal and transcript expression of CuZnSOD, CAT and monodehydroascorbate reductase (Romero-Puertas et al. 2007). Finally, a recent proteomic analysis of Cd exposed *Spinacea oleracea* plants showed differently modified profiles in apical and basal leaves (Fagioni and Zolla 2009). Also, metabolic analyses of Cd exposed spinach (*Spinacea oleracea*) plants showed reinforced 2-C-methyl-D-erythritol 2,4-cyclodiphosphate (MEcDP) accumulation, an intermediate of the methylerythritol 4-phosphate

(MEP) pathway, which was blocked (Rivasseau et al. 2009). Transcriptional responses of a *Chlamydomonas reinhardtii* cell wall-deficient mutant to metal stress was investigated and authors identified, sequenced, and quantified the induction of a number of transcripts that are up-regulated by cadmium chloride (Rubinelli et al. 2002).

2.2. CADMIUM GENOTOXICITY

The phenotype analysis of Cd exposed plants is of unquestionable importance, but in toxicology studies, phenotype studies should be complemented with analyses of the genome. This genome approach may allow a precocious detection of Cd effects and plant cell responses, allowing to survey putative occurrence of toxic effects, e.g. genotoxicity induced by the metal. Moreover, the combined phenotypic and molecular survey of Cd-exposed plants will allow its integration in prevention/intervention programs in the target agro-ecosystem.

Molecular and (cyto)genetic analyses are particularly important to assess putative genotoxicity as, like other metals, Cd can damage the genome or DNA of plants and/or induce mutations. There are different types of genotoxic effects: mutagenesis, which is a permanent change in DNA sequence within a gene; clastogenesis that refers to a damage in chromosome structure, usually resulting in a gain, loss or rearrangement of chromosome pieces within the genome; aneugenesis, which refers to the gain or loss of one or more chromosomes (aneuploidy) or to a complete haploid set of chromosomes (euploidy) (e.g. Panda and Panda 2002).

Genotoxicity induced by Cd has been extensively studied in mammals and particularly in humans. Moreover, exposure to this metal has been linked to several types of cancer, such as lung, prostate and renal cancer, and has been shown to induce tumours in experimental animals and exposed human cell lines (Waalkes 2003) and to induce large deletion mutations in mammalian cells (Filipic et al. 2006). Therefore, Cd and its compounds are classified as Category 1 human carcinogens by the International Agency for Research on Cancer (IARC 1993). However, the molecular mechanisms underlying the genotoxic and carcinogenic potential in organisms are still not well understood.

Cadmium may induce genotoxicity by inducing damages in DNA, and/or by leading to inactivation of mechanisms of DNA repair (Hartwig 1994). In this last model, Cd may interfere with DNA repair acting as a mutagen by direct inhibition of an essential DNA mismatch repair, resulting in a high level of genetic instability (Hartwig 1994, Jin et al. 2003, Slebos et al. 2006). Alternatively, genotoxicity may be induced indirectly by promoting the production of ROS, which may then damage nucleic acids (Hartwig 1994, Valverde et al. 2001, Apel and Hirt 2004). Furthermore other authors have shown that Cd cation can directly damage DNA, binding to DNA, possibly at guanine, adenine and thymine centres (Hossain and Huq 2002). Moreover, Cd may affect mitotic activity (often evaluated by the mitotic index (MI) and percentage of DNA synthesising cells (Fusconi et al. 2007).

Evaluation in plants of the Cd induced mutagenicity and/or genetic instability is of the utmost importance because, among several reasons, it may contribute to the inherited change of many phenotypic traits in the progeny of exposed plants. Furthermore, plants have been shown to provide ideal models for genotoxicity assays for screening as well as monitoring of environmental mutagens or genotoxins (Grant 1994, Knasmuller et al. 1998, Grant 1999). Different techniques have been used in plant bioassays for the detection of environmental metal pollution. A brief compilation of relevant works on molecular/genetic markers and their main achievements in the study of Cd genotoxicity in plants is presented in Table 1. Most of the studies on Cd genotoxicity used model species as *Arabidopsis* sp., *Vicia* sp., *Allium* sp. or *Oryza* sp. For example, Kovalchuk et al. (2001) developed an efficient and very sensitive system of transgenic *A. thaliana* that when exposed to several metals, one being Cd, exhibited a pronounced uptake-dependent increase in the frequencies of both somatic intrachromosomal recombination and point mutation.

Among the molecular markers, random-amplified polymorphic DNAs (RAPDs), amplified fragment length polymorphism (AFLP) and simple sequence repeats or microsatellite markers (SSR), open enormous advantages and need to be more explored in environmental risk assessments. Tanhuanpää et al (2007) compared four markers associated with grain Cd concentration in *Avena sativa*: 2 RAPDs, 1 REMAP (retrotransposon-microsatellite amplified polymorphism) and 1 SRAP (sequence-related amplified polymorphism), and as result suggested the use of molecular markers linked to differentiated Cd accumulation as an alternative to phenotyping selection.

Table 1. A brief review of genotoxicity studies in Cd exposed plants

Species studied	Method used	Genotoxicity (organ)	References
Allium cepa Vicia faba and Tradescantia	Micronucleus	Yes (root tips and pollen mother cells)	Steinkellner et al. (1998)
Bacopa monnieri	Comet assay	Yes (+ roots and - leaves)	Vaipayee et al. (2006)
Allium cepa	Chromosome aberrations	Yes (roots)	Borboa and de La Torre (1996)
Helianthus annuus	SSRs[3]	No (root and leaves)	Gomes et al. (2005)
Helianthus annuus	Flow cytometry	No (roots and leaves)	Azevedo (pers. comm.)
Hordeum vulgare	RAPDs[1]	Yes (root tips)	Liu et al. (2005)
Nicotiana tabacum	Comet assay	Yes (roots); No (leaves)	Gichner et al. (2004)
Oryza sativa	AFLP[2]	Yes (roots)	Aina et al. (2007)
Oryza sativa	RAPDs[1]	Yes (root tips)	Liu et al. (2007)
Phaseolus vulgaris	RAPDs[1]	Yes (seedlings)	Enan (2006)
Pisum sativum	Flow cytometry	Yes (roots)	Fusconi et al. (2006)
Thlaspi caerulescens and Thlaspi arvense	SSRs[3]	No (roots)	Paiva (2008)
Vicia faba	Micronucleus	Yes (root tips)	Beraud et al. (2007)
Vicia faba	Comet assay	Yes (leaves)	Lin et al. (2007)
Raphanus sativus L		Yeas roots (leaves and roots)	Villatoro-Pulido et al. (2009)
Allium sativum	Micronucleus	Yes (root tips)	Celik et al. 2008
Vicia faba	Micronucleus	Yes (root tips exposed to Leachates of municipal solid waste containing several metals)	Feng et al. 2009
Vicia faba	Comet assay	Yes (leaves)	Zhang et al. 2006
Bacopa monnieri L	Alkaline Comet assay	Yes (leaves and roots)	Vajpayee et al. 2006
Allium sativum and Vicia faba	Micronuclei	Yes (root tips)	Unyayar et al. 2006
Vicia faba	Comet assay	Yes (root tip)	Lin et al. 2005
Vicia faba	Micronucleus	Yes (root tips, exposed to municipoal solid waste with low concentrations of Cd)	Radetski et al. 2004
Tradescantia	Micronuclous	Yes (used as positive control)	Gong et al. 2003

Species studied	Method used	Genotoxicity (organ)	References
Allium cepa	Chromosome lagging of, multipolar anaphases, C-mitoses, etc	Yes (roots)	Dovgaliuk et al. 2001
Arabidopsis thaliana	Intrachromosomal recombination	Yes	Kovalchuk et al. 2001
Allium cepa	Anaphase-telophase chronmosome aberration	Yes (root)	Rank et al. 1998

[1] RAPDs - Random Amplified Polymorphic DNA.
[2] AFLP - Amplified Fragment Length Polymorphism.
[3] SSRs - Microsatellites or Simple Sequence Repeats.
[4] REMAP - retrotransposon-microsatellite amplified polymorphism).

Random Amplified Polymorphic DNA

These markers use random oligonucleotide (10-20) as initiators; have the advantage of being simple and non expensive, and if technical issues are adequately controlled, they may provide sensitive information on Cd induced genotoxicity. However, RAPD results are often difficult to reproduce around different laboratories (eg. Shasany et al. 2005). The low reproducibility is in fact considered the main disadvantage of this technique, namely when compared to others such as AFLPs.

Penner et al. (1995) identified RAPDs markers linked to a gene governing Cd uptake in durum wheat (*Triticum turgidum* L. var. *durum*). The genotoxicity induced by metals, including Cd, in kidney bean (*P. vulgaris*) seedlings was analysed by RAPDs, and revealed polymorphisms based on the presence and/or absence of DNA fragments in treated samples compared with untreated ones (Enan 2006). DNA polymorphisms detected by RAPDs analysis were also found in barley (*Hordeum vulgare*) (Liu et al. 2005 and 2009) and in rice (*Oryza sativa*) (Liu et al. 2007) exposed to Cd pollution, supporting the use of these markers in environmental toxicology.

Microsatellite or Single Sequence Repeat

SSRs are sequences of in general 1-6 nucleotide units repeated in tandem and randomly spread in plant genomes, usually in regions of low selective pressure. Deletions or duplications of those repeats through e.g. replication slippage, result in microsatellite instability (MSI) relative to other genomic loci. The variation in the repeat number is in nature so frequent that SSRs are very polymorphic and thus extremely useful for fine-scale genetic analysis, being used for genotyping and forensic analysis (GuhaMajumdar et al. 2008) and in the detection of genomic DNA damage and/or mutational events (e.g. deletions, insertions, point mutations). SSRs are likely to be one of the most reproducible techniques, especially when compared to RAPDs, which has the main disadvantage of low reproducibility with a consequent inconsistency of results (Powell et al. 1996, Jones et al. 1997). Due to these advantages, SSRs have been used to study genotoxic effects in several animal species (e.g. Zienolddiny et al. 2000, Jin et al. 2003, Ohshima 2003, Slebos et al. 2006). Despite providing robust information on MSI (putative mutations in the selected SSR), these markers are less used than other molecular markers, in particular when compared to RAPDs, in assessing metal induced genotoxicity. Some metals have been found to induce MSI in animal models: nickel (Ni) has been reported to promote genetic instability in hamster (Ohshima 2003) and in, e.g., human lung cancer cell lines (Zienolddiny et al. 2000), and exposure of human cell lines to environmentally relevant quantities of Cd led to statistically significant increases in MSI (e.g. Slebos et al. 2006).

In plant research, SSRs are already a powerful tool in e.g., environmental population genetics focusing on the relationships between environmental selective agents (stressors) and genotypic variability of plant natural populations (van Rossum et al. 2004). As for animals, despite SSRs have the potential to be used in the surveying of plant genomic DNA for evidence of genetic instability as in a genotoxic bioassay for the detection of DNA damage induced by environmental contaminants, up to moment, their use is limited and restricted to few reports. For instance, chloroplast SSR markers were efficient for elucidating the pattern of genetic differentiation in metal-tolerant populations in *Silene paradoxa* (Mengoni et al. 2001). For genomic DNA, using a population of 91 wheat (*T. aestivum* L.) recombinant inbred lines segregating for Al tolerance, Milla and Gustafson (2001) provided a genetic linkage map of the chromosome arm 4DL based on RFLP, SSR and AFLP markers. In another approach, van Rossum (2004) used five SSR loci to investigate the genetic structure in a metallicolous population of *A. halleri*.

Also, Ma et al. (2004) identified a SSR marker (Bmag353) associated with both Al resistance and citrate secretion, which provides a valuable tool for marker-assisted selection of Al-resistant lines. Several markers, among them SSR, were also used to map rye chromosome 7R and a new Alt gene (related with Al tolerance) was found to be located in this chromosome (Matos et al. 2005). In *T. caerulescens*, Basic and Besnard (2006) suggested the use of SSR markers for population genetic or mapping studies between and within several *Brassicaceae*, in particularly for genes involved in traits such as metal tolerance and/or hyperaccumulation. Similarly, Jimenez-Ambriz et al. (2007) compared neighbouring metallicolous and nonmetallicolous *T. caerulescens* populations growing in contaminated soils, using phenotypic and SSR markers. Recently, Monteiro et al. (2009a) demonstrated the occurrence of MSI and therefore genotoxicity in lettuce (*L. sativa*) plants exposed to Cd during 28 days since germination. Mutagenic effects of Cd were evaluated on nine SSR loci. No MSI was found in leaves, but a 2-bp deletion in one lettuce root SSR was detected among the SSRs that were analysed. However, this effect depends on the age/status of the plants and period of exposure, as the same authors also demonstrated that when five week old plants (not germinated in the presence of Cd) were exposed to Cd during 14 days under similar conditions as above, no genotoxicity was observed (Monteiro et al. 2005).

Amplified Fragment Length Polymorphism

This approach is based on DNA or cDNA (cDNA-AFLP) giving valuable information on putative point mutations in the genomes and/or functional information on genes controlling/associated with some traits, such as metal resistance. For the hyperaccumulator *A. halleri* ssp. *halleri*, Dräger et al. (2005) reported 128 transcript-derived sequence fragments (TDFs) identified in a cDNA-AFLP approach designed to identify metal-regulated transcripts in roots. In addition, the authors showed that in *A. halleri* roots the transcript levels of AhPDR11, encoding an ABC transport protein are induced in response to metal exposure. Fusco et al. (2005) also used cDNA-AFLP analysis to identify genes that exhibited a modulated expression following Cd treatment in *B. juncea*. Using AFLP markers and 13 co-dominant expressed sequence tags (EST), Deniau et al. (2006) constructed a map in *T. caerulescens*, used for mapping quantitative trit locus (QTL) for Zn and Cd concentrations in roots and shoots. Also, in the backcross segregating population produced from the crosses of *A. halleri* ssp. *halleri* (tolerant) and *A. lyrata* ssp. *petraea* (non tolerant), the analysis of the transcriptome by cDNA-amplified fragment length polymorphism (cDNA-AFLP) showed differently expressed transcripts according to different sensitivities to Cd (Cracium et al. 2006).

Most plant assays for genotoxicity focus however, cytogenetic techniques, in particular Comet assay and micronucleus (MNC), using model species as *Allium cepa*, *Vicia faba*, *Hordeum vulgare*, *Tradescantia sp.*, etc. These techniques, despite being highly valuable, are very time consuming, and their sensitivity may be hampered by technical constrain.

Comet Assay

This is a general but sensitive assay for genotoxicity, that measures DNA damage as percentage tail DNA (% tDNA), by using neutral or alkaline conditions that allow to detect single/double strand breakage. This technique has been used with success in several biological organisms (e.g. Juhel et al. 2007, Kuzmick et al. 2007) and even in human cells (e.g. Palus et al. 2003, Bakare et al. 2007). For instance, in plants, DNA damage measured by the Comet assay was performed in association with a battery of markers of oxidative stress (Lin et al. 2007), showing that oxidative stress induced by Cd accumulation in *Vicia faba* contributed to DNA damage.

Koppen et al. (1996) exposed *V. faba* roots to several toxic compounds, including Cd, and demonstrated that the Comet test was a suitable marker to assess genotoxicity, despite the heterogeneity in the extent of DNA migration reported. The Comet assay was evaluated as a potential tool for the assessment of ecogenotoxicity in a wetland plant, *Bacopa monnieri* L., exposed to Cd and the authors reported that this metal induced DNA damage, with more severe effects in roots than leaves (Vajpayee et al. 2006). Gichner et al. (2004) also compared Cd induced DNA damage in tobacco roots vs. leaves, showing through Comets that genotoxicity was only displayed in roots. Also, Zhang et al. (2006) assessed different types of DNA damage induced by Cd in leaves of *V. faba*. These authors found that Cd induced DNA damage (especially double strand breakages) together with apoptosis (detected by DAPI staining), revealing a possible relationship between apoptosis and DNA damage (Zhang et al. 2006).

Micronucleus Test

Due to mitotic anomalies and/or chromosome fragmentation, some chromosomes/fragments may not be incorporated into the daughter nucleus at cell division and lead to the formation of single or multiple micronuclei in the cytoplasm. These tests may be combined with a cytogenetic technique, the fluorescent in situ hybridization (FISH) that localizes (presence or absence) of specific chromosomic DNA sequences. For example, using a probe labeling the (peri-)centromeric region, FISH assay discriminates micronuclei containing a whole chromosome (labelling positive for the centromere) from micronuclei with acentric chromosome fragments (labelling negative for the centromere). Micronuclei test is however, time consuming and often subjective, and require cell division. Some criteria have to be followed, such as: the main nuclei and the micronuclei must be clearly separated, the diameter of the micronucleus should be approximately one third of the main nucleus, and should stain similarly.

Despite being time consuming, micronuclei tests have been largely used in toxicological assays for the last twenty years. Micronuclei frequency increased in water hyacinth (*Eichhornia crassipes*) exposed to Cd (Rosas et al. 1984). This data was confirmed by Mishra et al. (2007) who demonstrated that Cd levels affected the frequency of micronuclei and MI in root meristem of this species. Also it was observed an induction of micronuclei in *V. faba* root tips treated with Cd and Cr (De Marco et al. 1988). In *H. vulgare* roots, exposure to Cd increased mitotic irregularities comprising c-mitoses, anaphase bridges, breaks, stickiness, lagging and vagrant chromosomes and micronuclei (Zhang and Yang 1994). Also, in the same species, micronuclei increase was reported for *H. vulgare* roots exposed to Cd (Zhang and Xiao 1998). Similar results were described for *A. cepa* roots exposed to maleic hydrazide and methyl mercuric chloride, where the adaptive response after Cd treatment was also analysed (Panda et al. 1997). Dovgaliuk et al. (2001) showed for *A. cepa* apical meristem cells that Cd (and other metals) induced both clastogenic and aneugenic effects (including mitosis and cytokinesis disturbances). The toxicity of Cd in *V. faba* root tip cell was supported by the increase of micronuclei frequency, and this toxicity of Cd was increased by its combination with surfactant or acid rain (Liu et al. 2004). The increase of micronuclei induction, often concomitant with increases of oxidative stress parameters, was proposed to be interpreted as a consequence of oxidative stress, upholding the view that Cd-induced DNA damage is, to some extent, via generation of ROS (Rosa et al. 2003). Also, in *V. faba* exposed to Cd the population of micronuclei increased and a putative correlation with oxidative stress was proposed by the authors (Rosa et al. 2003). Ünyayar et al. (2006) tested different concentrations of Cd in two crop species, *A. sativum* and *V. faba*, and demonstrated that the frequency of micronuclei increased considerably for the concentration of 10 µM. Cadmium also increased micronuclei frequency in the medicinal species *Adhatoda vasica* (Jahangir et al. 2006). The exposure of *A. cepa* roots to several metals, including Cd, lowered the MI value and changed the proportion of mitotic phases (mainly prophases and telophases) in MI value. Mitotic disturbances (c-metaphases, sticky and lagging chromosomes, chromosome bridges, binucleate cells, micronuclei) were also observed (Glinska et al. 2007). Cadmium genotoxicity was assessed by micronuclei in *A. sativum*, and an induced frequency of micronuclei that exhibited a dose-dependent increase in Cd treatments was observed (Celik et al. 2008). Furthermore, this effect was suggested to be controlled by giberellic acid (Celik et al. 2008).

Flow Cytometry (FCM) Analyses

Initially, flow cytometry (FCM) has been used in health and biological research for many different purposes and appeared as a relatively rapid test applicable to any organism or tissue from which cellular or nuclear suspensions can be obtained. FCM has been less used in plant sciences, and most uses are essentially on nuclear DNA content (nDNA), ploidy and cell cycle analyses. FCM principle is based on the excitation by a laser of particles "cytomes" passing individually through a flow chamber in the flow cytometer. In response to this excitation each particle (cytome) scatters light (forward and side scatter) and/or emits fluorescence. For FCM fluorescence analyses, specific fluorochromes are used. For example for the analysis of DNA, propidium iodide (IP, intercalating) or DAPI (specific for AT bases) are used, allowing the analysis of the relative fluorescence of stained isolated nuclei. In plants, FCM analyses of relative nDNA content yields histograms showing a dominant peak (nuclei at the G_0/G_1 phase) and a minor peak (G_2 nuclei), separated by a low number of nuclei in S phase. To estimate ploidy levels, the position of the G_1 peak in a histogram of an unknown sample is compared to that of a reference plant with known ploidy. FCM presents several important advantages when compared to chromosome counting, namely: convenience (sample preparation is easy), rapid output (dozens of samples can be prepared and analysed in one working day), non requirement of dividing cells and it needs only few milligrams of tissue (e.g. Loureiro et al. 2006).

Flow cytometric assays detect minute differences in nDNA content and chromosomal damage produced by clastogenic agents through the quantification of the increase of the coefficient of variation (CV) of the G_0/G_1 peak (Otto and Oldiges 1980). Flow cytometry measurement of the dispersion in the nDNA content as induced by the interactions of DNA with environmental agents, emerged then as a powerful tool in cytogenetic investigations and in genotoxicity testing (Otto et al. 1981). It is largely used in animal ecotoxicological studies, for genotoxicity assessment (e.g. Bickham et al. 1998).

However, FCM utilization in plant ecotoxicological studies remain less explored. Changes in nDNA content were observed in *Zea mays* plants exposed to coal fly ash (McMurphy and Rayburn 1993) and to the fungicides captan (Rayburn et al. 1993) and triticonazole (Biradar et al. 1994). Moreover, the mean CV of G_1 peaks increased in *Z. mays* with coal fly ash treatments (McMurphy and Rayburn 1993), indicating DNA degradation. Using *Trifolium repens*, Citterio et al. (2002) proposed a new biomonitoring methodology for soil genotoxicity assessment, combining FCM with AFLP, and showed that DNA content changed (expressed as a decrease in DNA index) with increasing concentrations of metals, and increased the debris background at the highest concentrations of Cd and Cr tested. For exposure to polycyclic aromatic hydrocarbons, Aina et al. (2007) used the same method but did not found any differences in DNA content in exposed vs. control *T. repens* plants.

Monteiro et al. (2004 and 2007) exposed *in vivo* 5-week-old lettuce (*L. sativa*) plants during 14 days to a relatively high concentration of Cd (100 μM Cd) and assessed Cd induced genotoxicity by FCM. The authors found no changes in nDNA content or in the mean CV of the G_0/G_1 peak (a putative measure of DNA degradation). The same group also assessed genotoxic and mitogenic effects of Cd on the hyperaccumulator *T. caerulescens* vs. *Thlaspi arvense* L. (field pennycress) which is a non-accumulator plant, and found differences only for some organs. More recently, exposure of *P. sativum* plants to Cr (VI) revealed significant changes by the appearance of aneuploidies and polyploidization in the higher concentrations (Rodriguez et al. 2010). Moreover, rice or barley roots (but not leaves) from plants exposed to Al also showed changes in CV, and mitogenic effects (with accumulation of cells in one stage of the cell cycle).

Chapter 3

PLANT METAL ACCUMULATION AND HYPERACCUMULATION

3.1. ACCUMULATOR SPECIES

While the growth of most species is limited when growing on contaminated sites, some grow on contaminated sites and some are tolerant to toxic levels. When exposed to a highly contaminated environment, plants respond by excluding the metals (maintaining low and, within possible, constant metal concentration in their shoots up to a critical soil value) or accumulate the metal (accumulators) presenting therefore high levels of the metal inside the organs. As a particular case, indicator species have internal metal concentrations that reflect the external metal levels.

Although Cd is not an essential or beneficial element for plants, (hyper)accumulators generally exhibit measurable Cd concentrations, particularly in roots, but also in leaves, most probably as a result of inadvertent uptake and translocation (Assunção et al. 2003). A Cd foliar concentration above 100 µg/g DW (0.01%) is considered exceptional and it is used as a threshold value for Cd hyperaccumulation (100 mg/Kg DW) (Reeves and Baker 2000). The metal hyperaccumulation characteristic is not common in higher terrestrial plants and less than 0.2% of all angiosperms have been identified as metal hyperaccumulators (Reeves and Baker 2000).

Screening hyperaccumulators and accumulators is a key step, for both fundamental research on functional pathways involved in metal uptake-transport-accumulation processes, and for more applied studies concerning their use in metal contaminated soil phytoremediation.

Some species have been mostly studied in metal toxicology and phytoremediation studies, most of them due to their accumulation properties. Lettuce, an important human food crop, has been largely studied for the profile of Cd-distribution and cytotoxicity. Within the Brassicaceae (for review see Memon and Schroder 2009), in the genus *Thlaspi*, the alpine pennycress (*T. caerulescens*, Ganges ecotype) is a Zn, Cd and Ni hyperaccumulator plant commonly used as a model in metal transport and accumulation studies (e.g. Zhao et al. 2003).

Thlaspi caerulescens plants have been found by Reeves and Baker (2000) to contain more than 100 mg/Kg Cd frequently, and more than 1000 mg/Kg Cd occasionally, with very large variations between sites and populations, and considerable intrasite variability. Several studies have shown that *T. caerulescens* ecotype from metalliferous soils of a Zn/Pb mine spoil in the southern France (Ganges ecotype) is far superior in Cd accumulation than other ecotypes (e.g. Prayon from Belgium); in hydroponic conditions it was able to accumulate >10,000 mg/kg Cd in the shoots without showing any symptoms of phytotoxicity (Lombi et al. 2000). Recently it was proposed that Cd could play a physiological role in *T. caerulecens* by enhancing carbonic anhydrase activity (Liu et al. 2008); also the related non-accumulator, the field pennycress (*T. arvense* L.), is often used for comparative purposes.

Brassica juncea L. is also a well known Zn/Cd accumulator, in which the metal efflux transporter BjCET2 was demonstrated to play important roles in metal tolerance of this species (Xu et al. 2009). The high levels of Cd accumulation in its seeds (posing risks to human health) require studies on Cd redistribution in this species (e.g. Sankaran and Ebbs 2008).

Also field assays demonstrated that *Amaranthus retroflexus, Polygonum aviculare, Gundelia tournefortii, Noea mucronata* and *Scariola orientalis* accumulated metals, and could be used in phytorremediation strategies (Chehregani et al. 2009). Other species are known to accumulate metals, such as *Melastoma malabathricum* (Watanabe et al. 2008) and *Cassia siamea* (Jambhulkar et al. 2009).

Metal-Binding Ligands

Several plant species, namely accumulator and tolerant species, have evolved mechanisms to respond to metal uptake and accumulation that may include the chelation and sequestration of metals by particular ligands and, in some cases, the subsequent compartmentalization of the ligand-metal complex in vacuoles, achieving cellular metal homeostasis and consequent detoxification (e.g. Sung-Hyun et al. 2009).

There are many metal-binding biomolecules, most with role(s) in sequestering, transporting or storing the accumulated metal. Only a small fraction of the metal content in plants is present in the form of free aqua ions, and plants produce a number of possible ligands, including organic acids (e.g. carboxilic acids are present at high levels in plant vacuoles and play a role in hyperaccumulation, probably by sequestering them in the vacuole), nicotianamine (e.g. Gendre et al. 2007), amino acids (such as histidine), peptides and proteins (Damien et al. 2009).

Metallothioneins (MT) and PCs are particularly important in plant cell responses to metal exposure. MT are proteins of low molecular weight (approximately up 14000 Da). As MT are rich in cysteine, through the thiol group of its cysteine residues, they have the capacity to bind metals (either nutrient such as Zn, Cu) or xenobiotic (such as Cd or Hg). Concerning PCs, these are a family of enzymatically synthesized cysteine-rich peptides (Cobbett and Goldsbrough 2002). Damien et al. (2009) proposed a distinction between normal and hyperaccumulating plants, as nonaccumulating plants coordinate the majority of Cd and As using PCs and hyperaccumulators do not. Cadystins, Cd peptides, or γ-glutamyl peptides are nonprotein metal-binding polypeptides that possess the structure of (γGlu-Cys)nGly (Sung-Hyun et al. 2009).

Cadmium is recognised as a strong primary inducer of some of these metal binding ligands. Recently, Sung-Hyun et al. (2009) exposed *Echinochloa crusgalli* var. *frumentacea* to high Cd levels and purified a Cd-binding ligand (Cd-BL) from the roots and characterized its amino acid composition.

Plant vacuoles play crucial roles in the metal cellular homeostasis (Barkla and Pantoja 1996). The tonoplast functions as an effective and selective metal diffusion barrier. Vacuolar compartmentalization prevents the free circulation of Cd ions in the cytosol and forces them into a limited area (Sanitá di Toppi and Gabbrielli, 1999). Most progress on the role of the vacuole in metal transport, homeostasis and detoxification has been made with the micronutrients Zn and Fe (Martinoia et al. 2007). However, several studies also demonstrate the importance of the vacuole as a site of accumulation of a number of metals including Cd (Ma et al. 2005, Ueno et al. 2005). One example is the accumulation of Cd and PCs in the vacuole involving an ABC transporter (Hall 2002). Oat root tonoplast vesicles were found to accumulate Cd cation by a $2H^+$/ion antiport mechanism (Salt and Wagner 1993). Also, several transporters such as MTPs, CAXs and ABC transporters were already described as acting as Cd or Cd-chelate transporters (Martinoia et al. 2007).

Plant vacuoles of metal exposed plants accumulate PCs-Cd complexes in the vacuole (Cobbett and Goldsbrough 2002). This was evidenced, for example, by the preferential accumulation of Cd and/or PCs confined to the vacuole in tobacco mesophyll protoplasts derived from Cd-treated plants (Vogeli-Lange and Wagner 1990). Also, detoxification mechanisms were also described for *L. sativa* and *T. arvense* plants in which PCs play an important role (Ebbs et al. 2002, Maier et al. 2003). In *T. caerulescens* a main storage of Cd cation (as electron-dense granules) was found inside vacuoles by means of complexation with malate, an organic acid (Ma et al. 2005, Ueno et al. 2005).

Chapter 4

TROPHIC TRANSFER, BIOACCUMULATION AND BIOMAGNIFICATION OF CADMIUM

Plants have developed mechanisms for sequestering metals in their systems limiting metal phytotoxicity, but these plants may still be a threat to animal consumers and therefore a risk to ecological and human food chains (McLaughlin 2002). Animals and plants may amplify the Cd content relatively to their environment (Robards and Worsfold 1991, Greger 1999). Therefore, uptake of topsoil Cd (usually more contaminated that subsoil), and trophic transfer is likely the most important route for human exposure to Cd (IARC 1993). Cadmium bioaccumulation is affected by several factors such as its physico-chemical form, presence of other metals, pH, salinity, temperature, season, cation-exchange capacity of soils and the species taking up the Cd (e.g. Robards and Worsfold 1991). Several studies have demonstrated the transfer of Cd from producers, including plants: Xie et al. (2009) showed Cd bioaccumulation in mayfly fed on periphyton biofilms; Yu and Fleeger (2006) suggested that eutrophication by nitrate enrichment could enhance Cd trophic transfer from microalgae to suspension-feeding benthic invertebrates and Monteiro et al. (2008) demonstrated Cd accumulation in isopods fed with Cd biologically contaminated plants, including lettuce.

Metals biomagnification (progressive accumulation of a metal with increasing trophic levels) during transfer from contaminated soils may result in predatory exposure to toxic metal concentrations (Grenn and Tibbert 2008). Croteau and co-workers (2005) have demonstrated that Cd was progressively enriched among trophic levels in two discrete epiphyte-based food webs composed of macrophyte-dwelling invertebrates or fishes. Mathews and Fisher

(2008) have also shown biomagnification of Cd (and of MeHg and Po) from phytoplankton to zooplankton.

However, biomagnification process still remains a matter of controversy: trophic barriers were found along the soil-plant-herbivorous insect pathway reducing relative uptake of Cd (Merrington et al. 2001). Zhuang et al. (2009) demonstrated that transfer of metals (Pb, Zn Cu and Cd) to chicken from plant and insect was limited. Alonso et al. (2009) reported that a significant Cd transfer through a food chain (where *T. aestivum* was used as model plant in the chain) was found but not with the occurrence of biomagnification in the predator species (*Chrysoperla carnea*). Also, in a soil-plant arthropod system, Cd was biomagnified, contrarily to Zn (Green et al. 2006), in a soil-pea plant-aphid system, Zn was biomagnified while Cd was biominimized (Green and Tibbert 2008).

FACTORS AFFECTING TROPHIC TRANSFER OF METALS

The bioaccumulation of metals is known to differ among species and metals because of differences in uptake and loss rates, exposure pathways and influences of environmental parameters (Fisher and Reinfelder 1995, Wang and Fisher 1999). However, less is known about the influence of these factors in the internal storage and detoxification of accumulated metal and subsequent impacts on trophic transfer. Since the ingestion of metal-contaminated food can serve as a source of metals to consumers and can result in sub-lethal toxicity (e.g. Fisher and Hook 2002), understanding the mechanisms that influence metal trophic transfer is a critical step in the management of metal contaminated ecosystems. In general, to completely understand metal cycling through trophic levels, several factors which control the bioavailability of tissue-bound metals to predators must be considered and understood (e.g. tissue metal distributions and concentrations, duration of exposure, nutritional status and exposure history of predator). Different species will accumulate and partition metals in varying ways depending on the detoxification mechanisms employed. The subsequent bioavailability of those partitioned metals to a consumer will be dictated by digestive and assimilative mechanisms of its digestive tract and gut passage time (Wang and Fisher 1999). Added to this complexity is the varying ability of consumers to discriminate between different foods and contaminants, their nutritional status at the time of consumption, the degree of exposure, and the exposure history for the metal in question, all of which can influence the degree of metal assimilation (Wang and Fisher 1999).

SUBCELLULAR PARTITION OF METALS

The internal distribution and detoxification of metals within an organism can be used to explain trophic transfer of metals but also to predict metal toxicity for the organism itself. The internal metal sequestration strategies of different species are complex and variable and the determination of the metal concentrations in different compartments can be used to understand the complex relationship between metal accumulation and toxicity.

Metals can be present in various chemical forms in an organism, including the following: (a) free ionic form or complexed ion species (e.g., $CdCl_2$, $CdCl^+$, $CdCl_3^-$); (b) bound in the active center of functional proteins and enzymes; (c) bound to low molecular weight organic acids (e.g., citrate, malate); (d) bound to sequestration proteins (MT and PCs); (e) bound in vesicles of the lysosomal system, as intracellular granules; (f) precipitated in extracellular granules, mineral deposits, residual bodies, and exoskeletons; (g) bound to cellular constituents potentially causing dysfunction (e.g. DNA) (Vijver et al. 2004).

The various internal metal fractions all have their own binding capacity for metals, which has implications for food-chain transfer to higher trophic levels. A study on the relationship between subcellular Cd distribution in an oligochaete and its trophic transfer to a predatory shrimp showed that only metal present in the soluble fraction (organelles and protein fraction) of prey is available for the predator (Wallace et al. 1998). Factors influencing the subcellular distribution in the prey will directly alter trophic transfer to predators. Wallace et al. (1998) showed that differences in subcellular distribution of Cd between resistant and non-resistant worms directly affected Cd availability for the predatory shrimp. When fed resistant worms, shrimp absorbed about 4 times less Cd than when fed non-resistant worms (Wallace et al. 1998). Similar conclusions were found in a study using bivalves as prey, where the metal partitioning to organelles, denaturated cytosolic proteins, and MT comprise a subcellular compartment of that was considered as trophically available metal to predators (Wallace et al. 2003. Furthermore, Wallace and Luoma (2003) building on previous studies (Wallace and Lopez 1997, Wallace et al. 1998), postulated that Cd associated with the subcellular fractions, organelles, heat-denatured proteins, and heat-stable proteins of prey was trophically available metal and was assimilated at an efficiency of approximately 100% by the predator, while Cd bound to metal-rich granules was less bioavailable to predators. Such subcellular partitioning is dynamic in response to metal exposure and other environmental conditions, and is metal- and organism-specific. The different metal pools are not equally bioavailable to predators; thus, the determination of the metal concentration in the different subcellular compartments and the differences in its assimilation by consumers can be a useful tool to understand metal transfer to higher trophic levels.

Other applications of this approach have been proposed. Recent studies in aquatic organisms have revealed that the subcellular partitioning model may provide an improved method to predict Cd toxicity. As intracellular metal accumulation and the subsequent subcellular distribution of the metal in the cells are directly related to metal toxicity, it is likely that the metal concentration in a particular subcellular fraction will serve as a better toxicity predictor than the activity of the free metal ion in bulk solution (Wang and Rainbow 2006). Rainbow (2002) proposed that when accumulated metal destined for storage in a detoxified form (e.g. by MT and granules) exceeds the detoxified binding capacity, the metals are subsequently bound with other (metabolically available) forms, with the potential to cause toxicity to the organism. The significance of the subcellular distribution of accumulated metals in toxicity assessments is now receiving increasing attention among aquatic (e.g. Cheung et al. 2006, Perceval et al. 2006, Steen Redeker et al. 2007) and terrestrial organisms (Vijver et al. 2006, Vijver et al. 2007). Recently, Monteiro et al. (2008) demonstrated that Cd bound to heat-stable proteins was the least bioavailable fraction to isopods while Cd bound to heat-denatured proteins was the most trophically available to isopods, pointing out the ecological importance of the plant subcellular Cd distribution to the trophic transfer of Cd.

Chapter 5

CONCLUSIONS AND FUTURE PROSPECTS

Cadmium is one of the most highly toxic trace pollutants for humans, animals and plants. However, the major source of Cd contamination is due to anthropogenic activities. Several studies demonstrated its cytotoxic, mutagenic and/or carcinogenic effects in animal cells. In plants, available also data support that Cd leads to cytotoxic and genotoxic effects.

Cd interaction with plants and its toxicological effects at the plant physiological level, as well as factors conditioning its uptake by plants, are well described. In particular, several studies concern to Cd concentration in soil and its bioavailability is modulated by the presence of organic matter, pH, redox potential, temperature and concentrations of other elements. Once inside the plant, Cd usually leads to a battery of unspecific symptoms that include chlorosis, necrotic lesions, wilting, reddish coloration, growth reduction and disturbances in plant water relations. However, particular physiological aspects of Cd/plant interaction still remain obscure. At the carbon metabolism level, Cd inhibits photosynthetic pigments synthesis, electron transport, uptake of nutrients, CO_2 assimilation and enzymes of the Calvin cycle, and also induce a structural disorganization of chloroplasts. However, at what extent, directly or indirectly, this metal affects photosynthetic proteins and genes regulation is an open field for research. In this issue functional proteomics and, mostly, metabolomics (by detecting products) may contribute to elucidate Cd regulation of some related photosynthetic pathways. For example, the effects of cadmium on the transcription rate of chloroplast genes and/or the level of their transcripts remains poorly understood until now, although this is very important for deciphering the molecular mechanisms of cadmium damaging.

For animal cells it is well demonstrated that cadmium affects cellular processes such as cell-cycle progression, proliferation, differentiation, DNA replication and repair, and apoptosis, but the dimension of these effects in plant cells still remain unclear. It is known that Cd ions stimulate the free oxygen radical production and increased the activity of several antioxidant enzymes. Under severe and extreme stress conditions, the capacity of the antioxidant system may not be sufficient to minimize the harmful effect of oxidative injury. Plant survival under these conditions depends on its ability to perceive the stimulus, generate and transmit signals and induce biochemical changes. Research on signal molecules that mediate stress tolerance could be an important step towards a better understanding the capacity of plants to cope with these environments.

Together with other –omics, plant quantitative cytometry (also called plant cytomics) will bring new tools to better understand oxidative stress responses and how it may regulate, for example, gene expression of oxidative defence of other pathways such as cell cycle. Also, the expanding research using genomic technology, in particular DNA microarrays, allows a more comprehensive analysis of gene expression profiles of plants (from sensitive genomic model *A. thaliana* to the hyperaccumulator *T. caerulescens*) exposed to Cd. This will be a powerful tool to understand the expression of genes related to cellular protection and damage control mechanisms (e.g. genes coding for metallothioneins or antioxidant proteins) under Cd stress. In humans, as a whole, a number of genes appear to be coordinately regulated toward survival from Cd toxicity, and the evaluation of equivalent coordination networks in plants is still to clarify.

Concerning Cd induced-genotoxicity, most studies focus on genetic markers such as micronuclei or Comets. Molecular markers, despite less used, have in the last years demonstrated to be reliable markers in toxicity evaluations. Genetic methodologies (such as microsatellites, AFLP, RAPDs, micronucleus and flow cytometry) showed to be powerful tools to assay Cd genotoxicity. As specific markers of the genome characteristics/regions, one should emphasise the need to combine these markers for more comprehensive information. Moreover, the efficacy of these markers to verify genotoxic effects within an ERA perspective requires highly controlled experimental conditions and coordinated practices within the laboratories. Comparisons among these markers require controlled conditions (plant age, cadmium concentration and chemical formulation).

Accumulator and tolerant species have evolved mechanisms to respond to metal uptake and accumulation that include the chelation and sequestration of metals by particular ligands and, in some cases, the subsequent compartmentalization of the ligand-metal complex in vacuoles, achieving cellular metal homeostasis and consequent detoxification. Accumulator's species generally exhibit measurable Cd concentrations, particularly in roots, but also in leaves. Screening hyperaccumulators and accumulators could be the key step, for both fundamental research on functional pathways involved in metal uptake-transport-accumulation processes, to provide model systems to study the mechanisms of Cd tolerance and to use these species in phytoremediation strategies.

The use of soils and water contaminated with Cd in agriculture and the consume of contaminated products is especially important since it may represent a high risk to human food chain. Therefore, one concern of Cd pollution, like other heavy metals, is its trophic transfer, bioaccumulation and biomagnification. Bioaccumulation of metals differs among species and metals because of differences in uptake and loss rates, exposure pathways and influences of environmental parameter. Future investigation on the influence of these factors in the internal storage and detoxification of the accumulated metal and subsequent impacts on trophic transfer will improve the knowledge of metal bioaccumulation.

REFERENCES

2455/2001/EC European Commission Regulation (EC) no. 2455/2001 of 20 November 2001 established a list of priority substances in the field of water policy and amending Directive 2000/60/EC. *OJ L 331, p. 1–5.*

466/2001/EC, 2001. European Commission Regulation (EC) no. 466/2001 of 8 March 2001 setting maximum levels for certain contaminants in foodstuffs. 77: 1-13.

Aina, R., Labra, M., Fumagalli, P., Vannini, C., Marsoni, M., Cucchi, U., Bracale, M., Sgorbati, S., Citterio, S., 2007. Thiol-peptide level and proteomic changes in response to cadmium toxicity in *Oryza sativa* L. roots. *Environmental and Experimental Botany.* 59: 381-392.

Alonso, E., González-Núñez, M., Carbonell, G., Fernández, C., Tarazona, J.V., 2009. Bioaccumulation assessment via an adapted multi-species soil system (MS.3) and its application using cadmium. *Ecotoxicoloiy and Environmental Safety.* 72(4): 1038-44.

Apel, K., Hirt, H., 2004. Reactive oxygen species: metabolism, oxidative stress, and signal transduction. *Annual Review of Plant Biology.* 55: 373-399.

Aravind, P., Prasad, M.N., Malec, P., Waloszek, A., Strzałka, K., 2009. Zinc protects *Ceratophyllum demersum* L. (free-floating hydrophyte) against reactive oxygen species induced by cadmium. *Journal of Trace Elements in Medicine and Biology.* 23(1):50-60.

Assunção, A.G.L., Schat, H., Aarts, M.G.M., 2003. *Thlaspi caerulescens*, an attractive model species to study heavy metal hyperaccumulation in plants. *New Phytologist.* 159: 351-360.

ATSDR, 1999. Toxicological profile for cadmium. In: Agency for toxic substances and disease registry, draft for public comment. Public health service, U.S. department of health and human services, Atlanta, GA, pp. 1-397.

Azevedo, H., Pinto, G., Fernandes, J., Loureiro, J., Santos, C., 2005a. Cadmium effects on sunflower growth and photosynthesis. *Journal of Plant Nutrition.* 28, 2211-2220.

Azevedo, H., Pinto, G., Santos, C., 2005b. Cadmium effects in sunflower: Membrane permeability and changes in catalase and peroxidase activity in leaves and calluses. *Journal of Plant Nutrition.* 28: 2233-2241.

Azevedo, H., Pinto, G., Santos, C., 2005c. Cadmium effects in sunflower: Nutritional imbalances in plants and calluses. *Journal of Plant Nutrition.* 28: 2221-2231.

Bakare, A.A., Pandey, A.K., Bajpayee, M., Bhargav, D., Chowdhuri, D.K., Singh, K.P., Murthy, R.C., Dhawan, A., 2007. DNA damage induced in human peripheral blood lymphocytes by industrial solid waste and municipal sludge leachates. *Environmental and Molecular Mutagenesis.* 48(1): 30-7.

Balakhnina, T. Y., Kosobryukhov, A. A., Ivanov, A. A., Kreslavskii, V. D., 2005. The effect of cadmium on CO_2 exchange, variable fluorescence of chlorophyll, and the level of antioxidant enzymes in pea leaves. *Russian Journal of Plant Physiology.* 52(1): 15-20.

Ball, L., Accotto, G.P., Bechtold, U., Creissen, G., Funck, D., Jimenez, A., Kular, B., Leyland, N., Mejia-Carranza, J., Reynolds, H., Karpinski, S., Mullineaux, P.M., 2004. Evidence for a direct link between glutathione biosynthesis and stress defence gene expression in *Arabidopsis. The Plant Cell.* 16(9): 2448-62.

Barcelo, J., Poschenrieder, C., 1990. Plant water relations as affected by heavy metal stress: a review. *Journal of Plant Nutrition.* 13: 1-37.

Barkla, B.J., Pantoja, O., 1996. Physiology of ion transport across the tonoplast of higher plants. *Annual Review of Plant Physiology and Plant Molecular Biology.* 47: 159-184.

Baryla, A., Carrier, P., Franck, F., Coulomb, C., Sahut, C., Havaux, M., 2001. Leaf chlorosis in oilseed rape plants (*Brassica napus*) grown on cadmium-polluted soil: causes and consequences for photosynthesis and growth. *Planta.* 212: 696–709.

Basic, N., Besnard ,G., 2006. Gene polymorphisms for elucidating the genetic structure of the heavy-metal hyperaccumulating trait in *Thlaspi caerulescens* and their cross-genera amplification in *Brassicaceae. Journal of Plant Research.* 119(5): 479-87.

Ben, A., Charles, G., Hourmant, A., Ben, J., Branchard, M., 2009. Physiological behaviour of four rapeseed cultivar (*Brassica napus L.*) submitted to metal stress. *C. R. Biology.* 332(4): 363-70.

Benavides, M.P., Gallego, S.M. Tomaro, M.L., 2005. Cadmium toxicity in plants. *Brazilian Journal of Plant Physiology.* 17: 21-34.

Benzarti, S., Mohri, S., Ono, Y., 2008. Plant response to heavy metal toxicity: comparative study between the hyperaccumulator *Thlaspi caerulescens* (ecotype Ganges) and on accumulator plants: lettuce, radish, and alfalfa. *Environmental Toxicology.* 23(5): 607-16.

Beraud, E., Cotelle, S., Leroy, P. Ferard, J.F., 2007. Genotoxic effects and induction of phytochelatins in the presence of cadmium in *Vicia faba* roots. *Mutation Research.* 633: 112-116.

Bertin, G., Averbeck, D., 2006. Cadmium: cellular effects, modifications of biomolecules, modulation of DNA repair and genotoxic consequences (a review). Biochimie 88:1549–1559.

Bhartia, N., Singh, R.P., 1994. Antagonistic effect of sodium chloride to differential heavy metal toxicity regarding biomass accumulation and nitrate assimilation in *Sesamum indicum* seedlings. *Phytochemistry.* 35: 1157-1161.

Bi, Y., Chen, W., Zhang, W., Zhou, Q., Yun, L., Xing, D., 2009. Production of reactive oxygen species, impairment of photosynthetic function and dynamic changes in mitochondria are early events in cadmium-induced cell death in *Arabidopsis thaliana. Biology of the Cell.* 101(11): 629-43.

Bickham, J.W., Mazet, J.A., Blake, J., Smolen, M.J., Ballachey, B.E., 1998. Flow cytometric determination of genotoxic effects of exposure to petroleum in mink and sea otters. *Ecotoxicology.* 7: 191-199.

Biradar, D.P., Pedersen, W.L. Rayburn, A.L., 1994. Nuclear DNA analysis of maize seedlings treated with the triazole fungicide, triticonazole. Pesticides Science 41: 291-295.

Borboa, L., de La Torre, C., 1996. The genotoxicity of Zn(II) and Cd(II) in *Allium cepa* root meristematic cells. *New Phytologist.* 134: 481-486.

Burzyński, M., Kłobus, G., 2004. Changes of photosynthetic parameters in cucumber leaves under Cu, Cd, and Pb stress. *Photosynthetica.* 42 (4): 505-510.

Cai, S.W., Yue, L., Hu, Z.N., Zhong, X.Z., Ye, Z.L., Xu, H.D., Liu, Y.R., Ji, R.D., Zhang, W.H. Zhang, F.Y., 1990. Cadmium exposure and health effects among residents in an irrigation area with ore dressing wastewater. *Science of Total Environment.* 90: 67-73.

Campbell, P.G.C., 2006. Cadmium - A priority pollutant. *Environmental Chemistry*. 3: 387-388.

Carrier, P., Baryla, A., Havaux, M., 2003. Cadmium distribution and microlocalization in oilseed rape (*Brassica napus*) after long-term growth on cadmium-contaminated soil. *Planta*. 216: 939–950.

Cataldo, D.A., Garland, T.R. Wildung, R.E., 1983. Cadmium uptake kinetics in intact soybean plants. *Plant Physiology*. 73: 844-848.

Celik, A., Unyayar, S., Cekiç, F.O., Güzel, A., 2008. Micronucleus frequency and lipid peroxidation in *Allium sativum* root tip cells treated with gibberellic acid and cadmium. *Cell Biology and Toxicology*. 24(2): 159-64.

Chaffei, C., Pageau, K., Suzuki, A., Gouia, H., Ghorbel, H.M., Mascalaux-Daubresse, C., 2004. Cadmium toxicity induced changes in nitrogen management in *Lycopersicon esculentum* leading to a metabolic safeguard through an amino acid storage strategy. *Plant and Cell Physiology*. 45: 1681–1693.

Chaoui, A., El Ferjani, E., 2005. Effects of cadmium and copper on antioxidant capacities, lignification and auxin degradation in leave\s of pea (*Pisum sativum* L.) seedlings. *C. R. Biology*. 328: 23-31.

Chaoui, A., Mazhoudi, S., Ghorbal, M.H., El Ferjani, E., 1997. Cadmium and zinc induction of lipid peroxidation and effects on antioxidant enzyme activities in bean (*Phaseolus vulgaris* L.). *Plant Science*. 127: 139-147.

Chehregani, A., Noori, M., Yazdi, H.L., 2009. Phytoremediation of heavy-metal-polluted soils: screening for new accumulator plants in Angouran mine (Iran) and evaluation of removal ability. *Ecotoxicology and Environmental Safety*. 72(5): 1349-53.

Cheung, M.-S., Fok, E.M.W., Ng, T.Y.-T., Yen, Y.-F. Wang, W.-X., 2006. Subcellular cadmium distribution, accumulation, and toxicity in a predatory gastropod, *Thais clavigera*, fed different prey. *Environmental Toxicology and Chemistry*. 25:174-181.

Chiang, H.C., Lo, J.C., Yeh, K.C., 2006. Genes associated with heavy metal tolerance and accumulation in Zn/Cd hyperaccumulator *Arabidopsis halleri*: a genomic survey with cDNA microarray. *Environmental Science Technology*. 40(21): 6792-8.

Citterio, S., Aina, R., Labra, M., Ghiani, A., Fumagalli, P., Sgorbati, S. Santagostino, A., 2002. Soil genotoxicity assessment: a new strategy based on biomolecular tools and plant bioindicators. *Environmental Science Technology*. 36: 2748-2753.

Clemens, S., Antosiewicz, D.M., Ward, J.M., Schachtman, D.P., Schroeder, J.I., 1998. The plant cDNA LCT1 mediates the uptake of calcium and cadmium in yeast. *Proceedings of the National Academy of Sciences.* 95(20): 12043-8.

Clemens, S., Palmgren, M.G. Kramer, U., 2002. A long way ahead: understanding and engineering plant metal accumulation. *Trends in Plant Science.* 7: 309-315.

Cobbett, C., Goldsbrough, P., 2002. Phytocelatins and metallothioneins: roles in heavy metal detoxification and homeostasis. *Annual Review of Plant Biology.* 53: 159-182.

Costa, G., Morel, J.L., 1994. Efficiency of H^+-ATPase activity on cadmium uptake by 4 cultivars of lettuce. *Journal of Plant Nutrition.* 17: 627-637.

Craciun, A.R., Courbot, M., Bourgis, F., Salis, P., Saumitou-Laprade, P., Verbruggen, N., 2006. Comparative cDNA-AFLP analysis of Cd-tolerant and -sensitive genotypes derived from crosses between the Cd hyperaccumulator *Arabidopsis halleri* and *Arabidopsis lyrata* ssp. petraea. Journal of Experimental Botany 57(12): 2967-83.

Croteau, M.-N., Luoma, S.N. Stewart, A.R., 2005. Trophic transfer of metals along freshwater food webs: Evidence of cadmium biomagnification in nature Limnol. *Oceanography.* 50: 1511–1519.

Cseh, E., 2002. Metal permeability, transport and efflux in plants. In: Prasad, M. N. V., Strzalka, K., (Eds.). Physiology and Biochemistry of Metal Toxicity and Tolerance in Plants. Kluwer Academic Publishers, Dordrecht, Netherlands. pp. 1-36.

Dalla, Vecchia, F., La Rocca, N., Moro, I., de Faveri, S., Andreoli, C., Rascio, N., 2005. Morphogenetic, ultrastructural and physiological damages suffered by submerged leaves of *Elodea canadensis* exposed to cadmium. *Plant Science.* 168(2): 329–338.

Damien, L., Callahan, Alan, J. M., Baker, Spas, D., Kolev, Anthony, G., 2006. Wedd metal ion ligands in hyperaccumulating plants. Journal of *Biological* Inorganic Chemistry. 11: 2–12.

Delpérée, C., Lutts, L., 2008. Growth Inhibition Occurs Independently of Cell Mortality in Tomato (*Solanum lycopersicum*) Exposed to High Cadmium Concentrations. *Journal of Integrative Plant Biology.* 50(3): 300-310.

De Marco, A., Paglialunga, S., Rizzoni, M., Testa, A., Trinca, S., 1988. Induction of micronuclei in *Vicia faba* root tips treated with heavy metals (cadmium and chromium) in the presence of NTA. *Mutation Research.* 206(3): 311-5.

Deniau, A.X., Pieper, B., Ten, Bookum, W.M., Lindhout, P., Aarts, M.G., Schat, H., 2006. QTL analysis of cadmium and zinc accumulation in the heavy metal hyperaccumulator *Thlaspi caerulescens*. *Theoretical and Applied Genetics.* 113(5): 907-20.

Di Cagno, R., Guidi, L., De Gara, L., Soldatini, G., 2001. Combined cadmium and ozone treatments affect photosynthesis and ascorbate-dependent defences in sunflower. *New Phytologist.* 151 (3): 627-636(10).

Dietz, K.-J., Baier, M., Krämer, U., 1999. Free radicals and reactive oxygen species as mediators of heavy metal toxicity in plants. In: Prasad, M. N. V., Hagemeyer, J., (Eds.). *Heavy metal stress in plants.* Springer-Verlag, Berlin, Germany.

Dovgaliuk, A.I., Kaliniak, T.B., Blium, I.B., 2001. Cytogenetic effects of toxic metal salts on apical meristem cells of *Allium cepa* L. seed roots. *Toxicology Genetic.* 35(2):3-10.

Dräger, D.B., Voigt, K., Krämer, U., 2005. Short transcript-derived fragments from the metal hyperaccumulator model species *Arabidopsis halleri*. *Zeitschrift für Naturforschung* C. 60(3-4): 172-8.

Drazkiewicz, M., Skórzyńska-Polit, E., Krupa, Z., 2007. The redox state and activity of superoxide dismutase classes in *Arabidopsis thaliana* under cadmium or copper stress. *Chemosphere.* 67(1): 188-93.

Dunwei, C., Jiang, D., Dai, T., Jing, Q., Cao, W., 2009. Effects of cadmium on plant growth and physiological traits in contrast wheat recombinant inbred lines differing in cadmium tolerance. *Chemosphere.*

Ebbs, S., Lau, I., Ahner, B., Kochian, L., 2002. Phytochelatin synthesis is not responsible for Cd tolerance in the Zn/Cd hyperaccumulator *Thlaspi caerulescens* (J. and C. Presl). *Planta.* 214: 635-640.

Ekmekçi, Y., Tanyolaç, D., Ayhan, B., 2008. Effects of cadmium on antioxidant enzyme and photosynthetic activities in leaves of two maize cultivars. *Journal of Plant Physiology.* 165: 600-611.

Enan, M.R., 2006. Application of random amplified polymorphic DNA (RAPD) to detect the genotoxic effect of heavy metals. *Biotechnology Applied Biochemistry.* 43: 147-154.

Fagioni, M., Zolla, L., 2009. Does the different proteomic profile found in apical and basal leaves of spinach reveal a strategy of this plant toward cadmium pollution response? *Journal of Proteomic Research.* 8(5): 2519-29.

Feng, S., Wang, X., Wei, G., Peng, P., Yang, Y., Cao, Z., 2009. Leachates of municipal solid waste incineration bottom ash from Macao: heavy metal concentrations and genotoxicity. *Chemosphere.* 67(6):1133-7.

Filek, M., Keskinen, R., Hartikainen, H., Szarejko, I., Janiak, A., Miszalski, Z., Golda, A., 2008. The protective role of selenium in rape seedlings subjected to cadmium stress. *Journal of Plant Physiology.* 165(8): 833-44

Filipic, M., Fatur, T. Vudrag, M., 2006. Molecular mechanisms of cadmium induced mutagenicity. *Human Experimental Toxicology.* 25: 67-77.

Fisher, N.S. Hook, S.E., 2002. Toxicology tests with aquatic animals need to consider the trophic transfer of metals. *Toxicology.* 181: 531-536.

Fisher, N.S., Reinfelder, J.R., 1995. The trophic transfer of metals in marine systems. In: Tessier, A., Turner, D. R., (Eds.). *Metal Speciation and Bioavailability in Aquatic Systems.* John Wiley and Sons Ltd, London. pp. 363-406.

Fodor, F., 2002. Physiological responses of vascular plants to heavy metals. In: Prasad, M. N. V., Strzalka, K., (Eds.). Physiology and Biochemistry of Metal Toxicity and Tolerance in Plants. Kluwer Academic Publishers, Dordrecht, Netherlands. pp. 149-177.

Fusco, N., Micheletto, L., Dal Corso, G., Borgato, L., Furini, A., 2005. Identification of cadmium-regulated genes by cDNA-AFLP in the heavy metal accumulator *Brassica juncea* L. *Journal of Experimental Botany.* 56(421): 3017-27.

Fusconi, A., Gallo, C., Camusso, W., 2007. Effects of cadmium on root apical meristems of *Pisum sativum* L.: cell viability, cell proliferation and microtubule pattern as suitable markers for assessment of stress pollution. *Mutation Research.* 632(1-2): 9-19.

Fusconi, A., Repetto, O., Bona, E., Massa, N., Gallo, C., Dumas-Gaudot, E. Berta, G., 2006. Effects of cadmium on meristem activity and nucleus ploidy in roots of *Pisum sativum* L. cv. Frisson seedlings. *Environmental and Experimental Botany.* 58: 253-260.

Garnier, L., Simon-Plas, F., Thuleau, P., Agnel, J.P., Blein, J.P., Ranjeva, R., Montillet, J.L., 2006. Cadmium affects tobacco cells by a series of three waves of reactive oxygen species that contribute to cytotoxicity. *Plant Cell and Environment.* 29(10): 1956-69.

Gendre, D., Czernic, P., Conéjéro, G., Pianelli, K., Briat, J.F., Lebrun, M., Mari S., 2007. TcYSL3, a member of the YSL gene family from the hyperaccumulator *Thlaspi caerulescens*, encodes a nicotianamine-Ni/Fe transporter. *The Plant Journal.* 49(1):1-15.

Gichner, T., Patková, Z., Száková, J., Demnerová, K., 2004. Cadmium induces DNA damage in tobacco roots, but no DNA damage, somatic mutations or homologous recombination in tobacco leaves. *Mutation Research.* 559(1-2): 49-57.

Glińska, S., Bartczak, M., Oleksiak, S., Wolska, A., Gabara, B., Posmyk, M., Janas, K., 2007. Effects of anthocyanin-rich extract from red cabbage leaves on meristematic cells of *Allium cepa* L. roots treated with heavy metals. *Ecotoxicology and Environmental Safety.* 68(3): 343-50.

Gomes, A., Azevedo, H., Santos, C. Lopes, T., 2005. Microsatellite analysis of *Helianthus annuus* L genome exposed to cadmium and lead. 42nd Congress of Toxicology EUROTOX, Cracow, Poland.

Gomes-Junior, RA., Moldes, C.A., Delite, F.S., Pompeu, G.B., Gratão, P.L., Mazzafera, P., Lea, P.J., Azevedo, R.A., 2006. Antioxidant metabolism of coffee cell suspension cultures in response to cadmium. *Chemosphere.* 65:1330-1337.

Gong, P., Kuperman, R.G., Sunahara, G.I., 2003. Genotoxicity of 2,4- and 2,6-dinitrotoluene as measured by the Tradescantia micronucleus (Trad-MCN) bioassay. *Mutation Research.* 538(1-2):13-8.

Gong, J.-M., Lee, D.A., Schroeder, J.I., 2003. Long-distance root-to-shoot transport of phytochelatins and cadmium in *Arabidopsis*. *Proceedings of the National Academy of Sciences U. S. A.* 100: 10118-10123.

Grant, W.F., 1994. The present status of higher-plant bioassays for the detection of environmental mutagens. *Mutation Research.* 310: 175-185.

Grant, W.F., 1999. Higher plant assays for the detection of chromosomal aberrations and gene mutations - a brief historical background on their use for screening and monitoring environmental chemicals. *Mutation Research.* 426: 107-112.

Gratão, P.L., Monteiro, C.C., Rossi, M.L., Martinelli, A.P., Peres, L.E., Medici, L.O., Lea, P.J., Azevedo, R.A., 2009. Differential ultrastructural changes in tomato hormonal mutants exposed to cadmium. Environmental and Experimental Botany 67 (2): 387-394.

Green, I.D., Jeffries, C., Diaz, A., Tibbett, M. 2006. Contrasting behaviour of cadmium and zinc in a soil-plant-arthropod system. *Chemosphere.* 64(7): 1115-21.

Green, I.D., Tibbett, M., 2008. Differential uptake, partitioning and transfer of Cd and Zn in the soil-pea plant-aphid system. *Environmental Science and Technology.* 42(2): 450-5.

Greger, M., 1999. Metal availability and bioconcentration in plants. In: Prasad, M. N. V., Hagemeyer, J., (Eds.). Heavy metal stress in plants - from molecules to ecosystems. Springer-Verlag, Berlin, Germany. pp. 1-27.

GuhaMajumdar, M., Dawson-Baglien, E., Sears, B.B., 2008. Creation of a chloroplast microsatellite reporter for detection of replication slippage in *Chlamydomonas reinhardtii*. *Eukaryotic Cell.* 7(4): 639-46.

Hall, J.L., 2002. Cellular mechanisms for heavy metal detoxification and tolerance. *Journal of Experimental Botany.* 53: 1-11.

Han, S., Lee, J., Oh, C., Kim, P., 2006. Alleviation of Cd toxicity by composted sewage sludge in Cd-treated Schmidt birch (*Betula schmidtii*) seedlings. *Chemosphere.* 65 (4): 541-546.

Hartwig, A., 1994. Role of DNA-repair inhibition in lead-induced and cadmium-induced genotoxicity - a review. *Environmental Health Perspect.* 102: 45-50.

He, J., Ren, Y., Zhu, C., Yan, Y., Jiang, D., 2008. Effect of Cd on growth, photosynthetic gas exchange, and chlorophyll fluorescence of wild and Cd-sensitive mutant rice. *Photosynthetica.* 46: 466-470.

He, Q.B., Singh, B.R., 1993. Effect of organic-matter on the distribution, extractability and uptake of cadmium in soils. *Journal of Soil Science.* 44:641-650.

Heyno, E., Klose, C., Krieger-Liszkay, A., 2008. Origin of cadmium-induced reactive oxygen species production: mitochondrial electron transfer versus plasma membrane NADPH oxidase. *New Phytologist.* 179(3): 687-99.

Horemans, N., Raeymaekers, T., Van Beek, K., Nowocin, A., Blust, R., Broos, K., Cuypers, A., Vangronsveld, J., Guisez, Y., 2007. Dehydroascorbate uptake is impaired in the early response of *Arabidopsis* plant cell cultures to cadmium. *Journal of Experimental Botany.* 58(15-16): 4307-17.

Hossain, Z., Huq, F., 2002. Studies on the interaction between Cd^{2+} ions and nucleobases and nucleotides. *Journal of Inorganic Biochemistry.* 90: 97-105.

Huang, B., Kuo, S., Bembenek, R., 2003. Cadmium uptake by lettuce from soil amended with phosphorus and trace element fertilizers. *Water Air Soil Pollution.* 147: 109-127.

IARC, (1993). Beryllium, cadmium, mercury, and exposures in the glass manufacturing industry: *IARC Monogr. Eval. Carcinog Risks Hum.* Lyon, France

Jacoby, B., Moran, N., 2002. Mineral Nutrient Transport in Plants In. *Handbook of Plant Crop Physiology.* Marcel Dekker AG, Switzerland. pp 337-361.

Jahangir, T., Khan, T.H., Prasad, L., Sultana, S., 2006. Reversal of cadmium chloride-induced oxidative stress and genotoxicity by *Adhatoda vasica* extract in Swiss albino mice. *Biological Trace Element Research.* 111(1-3): 217-28.

Jambhulkar, H.P., Juwarkar, A.A., 2009. Assessment of bioaccumulation of heavy metals by different plant species grown on fly ash dump. *Ecotoxicology and Environmental Safety.* 72(4):1122-8.

Jiménez-Ambriz, G., Petit, C., Bourrié, I., Dubois, S., Olivieri, I., Ronce, O., 2007. Life history variation in the heavy metal tolerant plant *Thlaspi caerulescens* growing in a network of contaminated and non contaminated sites in southern France: role of gene flow, selection and phenotypic plasticity. *New Phytologist.* 173(1):199-215.

Jin, Y.H., Clark, A.B., Slebos, R.J.C., Al-Refai, H., Taylor, J.A., Kunkel, T.A., Resnick, M.A., Gordenin, D.A., 2003. Cadmium is a mutagen that acts by inhibiting mismatch repair. *Nature Genetics.* 34: 326-329.

John, D.A., Leventhal, J.S., 1996. Bioavailability of metals. In: du Bray, E. A., (Eds.). *Preliminary compilation of descriptive geoenvironmental mineral deposit models.* U.S. Department of the Interior, USA. pp. 10-18.

Jones, C.J., Edwards, K.J., Castaglione, S., Winfield, M.O., Sala, F., vandeWiel, C., Bredemeijer, G., Vosman, B., Matthes, M., Daly, A., Brettschneider, R., Bettini, P., Buiatti, M., Maestri, E., Malcevschi, A., Marmiroli, N., Aert, R., Volckaert, G., Rueda, J., Linacero, R., Vazquez, A. Karp, A., 1997. Reproducibility testing of RAPD, AFLP and SSR markers in plants by a network of European laboratories. *Molecular Breeding.* 3: 381-390.

Juhel, G., O'Halloran, J., Culloty, S.C., O'riordan, R.M., Davenport, J., O'Brien, N.M., James, K.F., Furey, A., Allis, O. 2007. In vivo exposure to microcystins induces DNA damage in the haemocytes of the zebra mussel, *Dreissena polymorpha,* as measured with the comet assay. *Environmental and Molecular Mutagenesis.* 48(1): 22-9.

Jung, M.C., Thornton, I., 1996. Heavy metal contamination of soils and plants in the vicinity of a lead-zinc mine, Korea. *Applied Geochemistry.* 11: 53-59.

Knasmuller, S., Gottmann, E., Steinkellner, H., Fomin, A., Pickl, C., Paschke, A., God, R., Kundi, M., 1998. Detection of genotoxic effects of heavy metal contaminated soils with plant bioassays. *Mutation Research.* 420: 37-48.

Koppen, G., Verschaeve, L., 1996. The alkaline comet test on plant cells: a new genotoxicity test for DNA strand breaks in Vicia faba root cells. *Mutation Research.* 360(3): 193-200.

Kovalchuk, O., Dubrova, Y.E., Arkhipov, A., Hohn, B.. Kovalchuk, I., 2000. Germline DNA - Wheat mutation rate after Chernobyl. *Nature.* 407: 583-584.

Krantev, A., Yordanova, R., Janda, T., Szalai, G., Popova, L., 2008. Treatment with salicylic acid decreases the effect of cadmium on photosynthesis in maize plants. *Journal of Plant Physiology.* 165: 920–931.

Krupa, Z., Moniak, M., 1998. The stage of leaf maturity implicates the response of the photosynthetic apparatus to cadmium toxicity. *Plant Science.* 138: 149-156.

Krupa, Z., Öquist, G., Huner, N.P.A., 1993. The effects of cadmium on photosynthesis of *Phaseolus vulgaris* - a fluorescence analysis. *Physiologia Plantarum.* 88: 626–630.

Krupa, Z., Siedlecka, A., Skórzynska-Polit Maksymiek, W., 2002. Heavy metal interactions with plant nutrients. In: Prasad, M. N. V., Strzalka, K., (Eds.). *Physiology and biochemistry of metal toxicity and tolerance in plants.* Kluwer Academic Publishers, Dordrecht, Netherlands. pp. 287-301.

Kuzmick, D.M., Mitchelmore, C.L., Hopkins, W.A., Rowe, C.L., 2007. Effects of coal combustion residues on survival, antioxidant potential, and genotoxicity resulting from full-lifecycle exposure of grass shrimp (*Palaemonetes pugio* Holthius). *Science Total Environment.* 373(1): 420-30.

Lane, T.W., Morel, F.O.M.M., 2000. A biological function for cadmium in marine diatoms. *Proceedings of the National Academy of Sciences U. S. A.* 97: 4627-4631.

Lane, T.W., Saito, M.A., George, G.N., Pickering, I.J., Prince, R.C. Morel, F.M.M., 2005. Biochemistry: A cadmium enzyme from a marine diatom. *Nature.* 435: 42-42.

Larbi, A., Morales, F., Abadía, A., Gogorcena, Y., Lucena, J.J., Abadía, J., 2002. Effects of Cd and Pb in sugar beet plants grown in nutrient solution: induced Fe deficiency and growth inhibition. *Functional Plant Biology.* 29: 1453–1464.

Lee, S., Ahsan, N., Lee, K.W., Kim, D.H., Lee, D.G., Kwak, S.S., Kwon, S.Y., Kim, T.H., Lee, B.H., 2007. Simultaneous overexpression of both CuZn superoxide dismutase and ascorbate peroxidase in transgenic tall fescue plants confers increased tolerance to a wide range of abiotic stresses. *Journal of Plant Physiology.* 164(12): 1626-38.

Li, L., He, Z., Pandey, G.K., Tsuchiya, T., Luan, S., 2002. Functional cloning and characterization of a plant efflux carrier for multidrug and heavy metal detoxification. *Journal of Biology and Chemistry.* 277(7):5360-8.

Lin, A.J., Zhang, X.H., Chen, M.M., Cao, Q., 2007. Oxidative stress and DNA damages induced by cadmium accumulation. *Journal of Environmental Science (China)*. 19(5): 596-602.

Lin, A.J., Zhu, Y.G., Tong, Y.P., Geng, C.N., 2005. Evaluation of genotoxicity of combined pollution by cadmium and atrazine. *Bulletin of Environmental Contamination and Toxicology*. 74(3): 589-96.

Linger, P., Ostwald, A., Haensler, J., 2005. *Cannabis sativa* L. growing on heavy metal contaminated soil: growth, cadmium uptake and photosynthesis. *Biologia Plantarum*. 49: 567-576.

Liu, H., Liao, B., Lu, S., 2004. Toxicity of surfactant, acid rain and Cd2+ combined pollution to the nucleus of *Vicia faba* root tip cells. *Ying Yong Sheng Tai Xue Bao*. 15(3):493-6.

Liu, J., Qian, M., Cai, G., Zhu, Q., Wong, M.H., 2007. Variations between rice cultivars in root secretion of organic acids and the relationship with plant cadmium uptake. *Environmental Geochemistry and Health*. 29(3):189-95.

Liu, M.Q., Yanai, J., Jiang, R.F., Zhang, F., McGrath, S.P., Zhao, F.J., 2008. Does cadmium play a physiological role in the hyperaccumulator *Thlaspi caerulescens*? *Chemosphere*. 71(7):1276-83.

Liu, W., Li, P.J., Qi, X.M., Zhou, Q.X., Zheng, L., Sun, T.H., Yang, Y.S., 2005. DNA changes in barley (*Hordeum vulgare*) seedlings induced by cadmium pollution using RAPD analysis. *Chemosphere*. 61(2):158-67.

Liu, W., Yang, Y.S., Li, P.J., Zhou, Q.X., Xie, L.J., Han, Y.P. 2009. Risk assessment of cadmium-contaminated soil on plant DNA damage using RAPD and physiological indices. *Journal of Hazardous Material*. 161(2-3):878-83.

Lombi, E., Zhao, F.J., Dunham, S.J. McGrath, S.P., 2000. Cadmium accumulation in populations of *Thlaspi caerulescens* and *Thlaspi goesingense*. *New Phytologist*. 145: 11-20.

López-Millán, A-F., Sagardoy, R., Solanas, M., Abadía, A., Abadía, J., 2009. Cadmium toxicity in tomato (*Lycopersicon esculentum*) plants grown in hydroponics. *Environmental and Experimental Botany*. 65: 376–385.

Loureiro, J., Rodriguez, E., Doležel, J., Santos, C., 2006. Flow cytometric and microscopic analysis of the effect of tannic acid on plant nuclei and estimation of DNA content. *Annals of Botany*. 98:515–527.

Ma, J.F., Nagao, S., Sato, K., Ito, H., Furukawa, J., Takeda, K., 2004. Molecular mapping of a gene responsible for Al-activated secretion of citrate in barley. *Journal of Experimental Botany*. (401):1335-41.

Ma, J.F., Ueno, D., Zhao, F.J., McGrath, S.P., 2005. Subcellular localisation of Cd and Zn in the leaves of a Cd-hyperaccumulating ecotype of *Thlaspi caerulescens*. *Planta.* 220: 731-736.

Maier, E.A., Matthews, R.D., McDowell, J.A., Walden, R.R., Ahner, B.A., 2003. Environmental cadmium levels increase phytochelatin and glutathione in lettuce grown in a chelator-buffered nutrient solution. *Journal of Environmental Quality.* 32: 1356-1364.

Malik, D., Sheoran, I.S., Singh, R., 1992. Carbon metabolism in leaves of cadmium treated wheat seedlings. *Plant Physiology and Biochemistry.* 30: 223-229.

Mallick, N., Mohn, F.H., 2003. Use of chlorophyll fluorescence in metal-stress research: a case study with green microalga *Scenedesmus*. *Ecotoxicology Environment and Safety.* 55:64–9.

Martnoia, E., Maeshima, M., Neuhaus, H.E., 2007. Vacuolar transporters and their essential role in plant metabolism. *Jounal of Experimental Botany.* 58(1): 83-102.

Mathews, T., Fisher, N.S., 2008. Evaluating the trophic transfer of cadmium, polonium, and methylmercury in an estuarine food chain. *Environmental Toxicology and Chemistry.* 27(5):1093-101.

Matos, M., Camacho, M.V., Pérez-Flores, V., Pernaute, B., Pinto-Carnide, O., Benito, C., 2005. A new aluminum tolerance gene located on rye chromosome arm 7RS. *Theoretical and Applied Genetics.* 111(2): 360-9.

McLaughlin, M.J., 2002. Bioavailability of metals to terrestrial plants. In: Allen, H. E., (Eds.). Bioavailability of metals in terrestrial ecosystems: importance of partitioning for bioavailability to invertebrates, microbes, and plants. Society of Environmental Toxicology and Chemistry (SETAC), Lawrence, USA. pp. 39-68.

McLaughlin, M.J., Whatmuff, M., Warne, M., Heemsbergen, D., Barry, G., Bell, M., Nash, D., Pritchard, D., 2006. A field investigation of solubility and food chain accumulation of biosolid-cadmium across diverse soil types. *Environmental Chemistry.* 3: 428-432.

McMurphy, L.M., Rayburn, A.L., 1993. Nuclear alterations of maize plants grown in soil contaminated with coal fly ash. *Archives of Environmental Contamination and Toxicology.* 25: 520-524.

Memon, A.R., Schröder, P., 2009. Implications of metal accumulation mechanisms to phytoremediation. Environmental Science and Pollution *Research International.* 16(2):162-75.

Mengoni, A., Barabesi, C., Gonnelli, C., Galardi, F., Gabbrielli, R. Bazzicalupo, M., 2001. Genetic diversity of heavy metal-tolerant populations in *Silene paradoxa* L. (*Caryophyllaceae*): a chloroplast microsatellite analysis. *Molecular Ecology.* 10: 1909-1916.

Merrington, G., Miller, D., McLaughlin, M.J., Keller, M.A., 2001. Trophic barriers to fertilizer Cd bioaccumulation through the food chain: a case study using a plant--insect predator pathway. *Archives of Environmental Contamination and Toxicology.* 41(2):151-6.

Milla, R., Gustafson, J.P., 2001. Genetic and physical characterization of chromosome 4DL in wheat. *Genome.* 44(5): 883-92.

Mishra, K.K., Rai, U.N., Prakash, O., 2007. Bioconcentration and phytotoxicity of Cd in *Eichhornia crassipes*. *Environmental Monitoring and Assessment.* 130(1-3): 237-43.

Mittler, R., 2002. Oxidative stress, antioxidants and stress tolerance. *Trends in Plant Science.* 7: 405-410.

Mobin, M., Khan, N.A., 2007. Photosynthetic activity, pigment composition and antioxidative response of two mustard (*Brassica juncea*) cultivars differing in photosynthetic capacity subjected to cadmium stress. *Journal of Plant Physiology.* 164: 601-10.

Monteiro, M.S., Lopes, T., Loureiro, J., Mann, R.M., Santos, C., Soares, A.M.V.M., 2004. An assessment of genetic stability in cadmium-treated *Lactuca sativa* L. by flow cytometry and microsatellite analysis. Fourth SETAC World Congress and 25th Annual Meeting in North America, Portland, Oregon, USA.

Monteiro, M.S., Lopes, T., Mann, R.M., Paiva, C., Soares, A.M., Santos, C., 2009a. Microsatellite instability in *Lactuca sativa* chronically exposed to cadmium. *Mutation Research.* 672(2): 90-4.

Monteiro, M., Santos, C., Mann, R.M., Soares, A.M.V.M., Lopes, T., 2007. Evaluation of cadmium genotoxicity in *Lactuca sativa* L. using nuclear microsatellites. *Environmental and Experimental Botany.* 60: 421-427.

Monteiro, M.S., Rodriguez, E., Loureiro, J., Mann, R.M., Soares, A., Santos, C., 2005. The use of flow cytometry to assess the ploidy stability of cadmium-treated *Lactuca sativa* L.: a potential use of this technique for toxicological studies. Congresso da Sociedade Ibérica de Citometria, Porto, Portugal.

Monteiro, M.S., Santos, C., Soares, A.M., Mann, R.M., 2008. Does subcellular distribution in plants dictate the trophic bioavailability of cadmium to *Porcellio dilatatus* (Crustacea, Isopoda)? *Environmental Toxicology and Chemistry.* 27(12): 2548-56.

Monteiro, M.S., Santos, C., Soares, A.M., Mann, R.M., 2009b. Assessment of biomarkers of cadmium stress in lettuce. *Environmental Toxicology and Chemistry.* 72(3): 811-8.

Mysliwa-Kurdziel, B., Strzalka, K., 2002. Influence of metal on biosynthesis of photosynthetic pigments. In: Prasad, M. N. V., Strzalka, K., (Eds.). Physiology and biochemistry of metal toxicity and tolerance in plants. Kluwer Academic Publishers, Dordrecht, Netherlands. pp. 201-227.

Nelson, W.O., Campbell, P.G.C., 1991. The effects of acidification on the geochemistry of Al, Cd, Pb and Hg in freshwater environments: A literature review. *Environmental Pollution.* 71: 91-130.

Nogawa, K., Honda, R., Kido, T., Tsuritani, I., Yamada, Y., Ishizaki, M. Yamaya, H., 1989. A dose-response analysis of cadmium in the general environment with special reference to total cadmium intake limit. *Environmental Research.* 48: 7-16.

OECD, 2006. Terrestrial plant test: seedling emergence and seedling growth test. OECD-Organisation for Economic Co-operationand Development, Paris, pp 208–221.

Ohshima, S., 2003. Induction of genetic instability and chromo nickel sulfate in V79 Chinese hamster cells. *Mutagenesis.* 18: 133-137.

Ortega-Villasante, C., Hernández, L.E., Rellán-Alvarez, R., Del Campo, F.F., Carpena-Ruiz, R.O., 2007. Rapid alteration of cellular redox homeostasis upon exposure to cadmium and mercury in alfalfa seedlings. *New Phytologist.* 176(1): 96-107.

Otto, F.J., Oldiges, H., 1980. Flow cytogenetic studies in chromosomes and whole cells for the detection of clastogenic effects. *Cytometry.* 1: 13-17.

Otto, F.J., Oldiges, H., Göhde, W., Jain, V.K., 1981. Flow cytometric measurement of nuclear DNA content variations as a potential in vivo mutagenicity test. *Cytometry.* 2: 189 – 191.

Ouariti, O., Boussama, N., Zarrouk, M., Cherif, A., Ghorbal, M.H., 1997. Cadmium - and copper-induced changes in tomato membrane lipids. *Phytochemistry.* 45: 1343-1350.

Paiva, C., 2008. Avaliação da genotoxicidade do cádmio em duas espécies de *Thlaspi*. MSc Thesis, Universidade de Aveiro, Aveiro, pp. 59.

Pal'ove-Balang, A., Kisová, J., Pavlovkin, I., Mistrík, 2006. Effect of manganese on cadmium toxicity in maize seedlings. *Plant Soil Environment.* 5(4): 143–149.

Palus, J., Rydzynski, K., Dziubaltowska, E., Wyszynska, K., Natarajan, A.T., Nilsson, R., 2003. Genotoxic effects of occupational exposure to lead and cadmium. *Mutation Research.* 540(1): 19-28.

Panda, B.B., Panda, K.K., 2002. Genotoxicity and mutagenicity of metals in plants. In: Prasad, M.N.V., Strzalka, K., (Eds.). Physiology and Biochemistry of Metal Toxicity and Tolerance in Plants. Kluwer Academic Publishers, Dordrecht, Netherlands. pp. 395-414.

Panda, K.K., Patra, J., Panda, B.B., 1997. Persistence of cadmium-induced adaptive response to genotoxicity of maleic hydrazide and methyl mercuric chloride in root meristem cells of *Allium cepa* L.: differential inhibition by cycloheximide and buthionine sulfoximine. *Mutation Research.* 389(2-3): 129-39.

Penner, G.A., Bezte, L.J., Leisle, D., Clarke, J., 1995. Identification of RAPD markers linked to a gene governing cadmium uptake in durum wheat. *Genome.* 38(3): 543-7.

Perceval, O., Couillard, Y., Pinel-Alloul, B. Campbell, P.G.C., 2006. Linking changes in subcellular cadmium distribution to growth and mortality rates in transplanted freshwater bivalves (*Pyganodon grandis*). *Aquatic Toxicology.* 79: 87-98.

Peris, M., Mico, C., Recatala, L., Sanchez, R., Sanchez, J., 2007. Heavy metal contents in horticultural crops of a representative area of the European Mediterranean region. *Science of Total Environment.* 378: 42-48.

Pietrini, F., Iannelli, M.A., Pasqualini, S., Massacci A., 2003. Interaction of Cadmium with Glutathione and Photosynthesis in Developing Leaves and Chloroplasts of *Phragmites australis* (Cav.) Trin. ex Steudel. *Plant Physiology.* 133: 829–837.

Powell, W., Morgante, M., Andre, C., Hanafey, M., Vogel, J., Tingey, S., Rafalski, A., 1996. The comparison of RFLP, RAPD, AFLP and SSR (microsatellite) markers for germplasm analysis. *Molecular Breeding.* 2: 225-238.

Prasad, M.N.V., 1995. Cadmium toxicity and tolerance in vascular plants. *Environmental and Experimental Botany.* 35: 525-545.

Radetski, C.M., Ferrari, B., Cotelle, S., Masfaraud, J.F., Ferard, J.F., 2004. Evaluation of the genotoxic, mutagenic and oxidant stress potentials of municipal solid waste incinerator bottom ash leachates. *Science of Total Environment.* 333(1-3): 209-16.

Rainbow, P.S., 2002. Trace metal concentrations in aquatic invertebrates: why and so what? *Environmental Pollution.* 120: 497-507.

Rank, J., Nielsen, M.H., 1998 .Genotoxicity testing of wastewater sludge using the *Allium cepa* anaphase-telophase chromosome aberration assay. *Mutation Research.* 418(2-3): 113-9.

Rayburn, A.L., Pedersen, W.L., McMurphy, L.M., 1993. The fungicide captan reduces nuclear DNA content in maize seedlings. *Pesticide Science.* 37: 79-82.

Razinger, J., Drinovec, L., Zrimec, A., 2009. Real-time visualization of oxidative stress in a floating macrophyte *Lemna minor* L. exposed to cadmium, copper, menadione, and AAPH. *Environmental Toxicology.* 23.

Reeves, R.D., Baker, A.J.M., 2000. Metal-accumulating plants. In Raskin, I., Ensley, B.D. (Eds.). *Phytoremediation of Toxic Metals: Using Plants to Clean up the Environment.* John Wiley and Sons Inc., New York, NY, USA. pp. 193–229.

Rivasseau, C., Seemann, M., Boisson, A.M., Streb, P., Gout, E., Douce, R., Rohmer, M., Bligny, R., 2009. Accumulation of 2-C-methyl-D-erythritol 2,4-cyclodiphosphate in illuminated plant leaves at supraoptimal temperatures reveals a bottleneck of the prokaryotic methylerythritol 4-phosphate pathway of isoprenoid biosynthesis. *Plant Cell Environment.* 32(1):82-92.

Robards, K., Worsfold, P., 1991. Cadmium: toxicology and analysis - a review. *Analyst.* 116: 549-568.

Rodriguez, E., Fernandes, P., Azevedo, R., Fernandes, J., Araújo, A., Santos, C., 2010. Genotoxic and cytotoxic evaluation of *Pisum sativum* plants exposed to Chromium. *Europe 20th Annual Meeting of the Society of Environmental Toxicology and Chemistry* (SETAC). Sevilla.

Rodríguez-Serrano, M., Romero-Puertas, M.C., Pazmiño, D.M., Testillano, P.S., Risueño, M.C., Del Río, L.A., Sandalio, L.M., 2009. Cellular response of pea plants to cadmium toxicity: cross talk between reactive oxygen species, nitric oxide, and calcium. *Plant Physiology.* 150(1): 229-43.

Rodríguez-Serrano, M., Romero-Puertas, M.C., Zabalza, A., Corpas, F.J., Gómez, M., Del Río, L.A., Sandalio, L.M., 2006. Cadmium effect on oxidative metabolism of pea (*Pisum sativum* L.) roots. Imaging of reactive oxygen species and nitric oxide accumulation in vivo. *Plant Cell and Environment.* 29(8): 1532-44.

Romero-Puertas, M.C., Corpas, F.J., Rodríguez-Serrano, M., Gómez, M., Del Río L.A., Sandalio, L.M., 2007. Differential expression and regulation of antioxidative enzymes by cadmium in pea plants. *Journal of Plant Physiology.* 164(10): 1346-57.

Rosa, E.V., Valgas, C., Souza-Sierra, M.M., Corrêa, A.X., Radetski, C.M., 2003. Biomass growth, micronucleus induction, and antioxidant stress enzyme responses in *Vicia faba* exposed to cadmium in solution. *Environmental Toxicology and Chemistry.* 22(3): 645-9.

Rosas, I., Carbajal, M.E., Gómez-Arroyo, S., Belmont, R., Villalobos-Pietrini, R., 1984. Cytogenetic effects of cadmium accumulation on water hyacinth (*Eichhornia crassipes*). *Environmental Research.* 33(2):386-95.

Rubinelli, P., Siripornadulsil, S., Gao-Rubinelli, F., Sayre, R.T., 2002. Cadmium- and iron-stress-inducible gene expression in the green alga *Chlamydomonas reinhardtii*: evidence for H43 protein function in iron assimilation. *Planta.* 215(1):1-13.

Salt, D.E., Wagner, G.J., 1993. Cadmium transport across tonoplast of vesicles from oat roots. Evidence for a Cd^{2+}/H^+ antiport activity. *Journal of Biology and Chemistry.* 268: 12297-12302.

Sandalio, L., Dalurzo, H., Gomes, M., Romero-Puertas, M., del Rio, L., 2001. Cadmium-induced changes in the growth and oxidative metabolism of pea plants. *Journal of Experimental Botanny.* 52: 2115–2126.

Sanitá di Toppi, L., Gabbrielli, R., 1999. Response to cadmium in higher plants. *Environmental and Experimental Botany.* 41: 105-130.

Sankaran, R.P., Ebbs, S.D., 2008. Transport of Cd and Zn to seeds of Indian mustard (*Brassica juncea*) during specific stages of plant growth and development. *Physiologia Plantarum.* 132(1):69-78.

Schützendübel, A., Polle, A., 2002. Plant responses to abiotic stresses: heavy metal-induced oxidative stress and protection by mycorrhization. *Journal of Experimental Botany.* 53: 1351-1365.

Shasany, A. K., Darokar, M. P., Dhawan, S., Gupta, A. K., Gupta, S., Shukla, A. K., Patra, N. K., Khanuja, S. P. S. 2005. Use of RAPD and AFLP Markers to Identify Inter- and Intraspecific Hybrids of *Mentha*. *Journal of Heredity Advance Access.* 96: 542-549.

Sheoran, I.S., Singal, H.R., Singh, R., 1990. Effect of cadmium and nickel on photosynthesis and enzymes of the photosynthetic carbon reduction cycle in pigeon pea (*Cajanus cajan* L.). *Photosynthesis Research.* 23: 345–351.

Shigaki, T., Barkla, B.J., Miranda-Vergara, M.C., Zhao, J., Pantoja, O., Hirschi, K.D.J., 2005. Identification of a crucial histidine involved in metal transport activity in the *Arabidopsis* cation/H+ exchanger CAX1. *Chemical Biology.* 280(34):30136-42.

Siedlecka, A., Baszynski, T., 1993. Inhibition of electron flow around photosystem I in chloroplasts of Cd-treated maize is due to Cd-induced iron deficiency. *Physiologia Plantarum.* 87: 199–202.

Siedlecka, A. Krupa, Z., 1996. Interaction between cadmium and iron and its effects on photosynthetic capacity of primary leaves of *Phaseolus vulgaris*. *Plant Physiology and Biochemistry.* 34: 833-841.

Siedlecka, A., Krupa, Z., Samuelsson, G., Öquist, G., Gardeström, P., 1997. Primary carbon metabolism in *Phaseolus vulgaris* plants under Cd/Fe interaction. *Plant Physiology and Biochemistry.* 35: 951–957.

Singh, P.K., Tewari, R.K., 2003. Cadmium toxicity induced changes in plant water relations and oxidative metabolism of *Brassica juncea* L. plants. *Journal of Environmental Biology.* 24(1):107-12.

Singh, R.P., Agrawal, M., 2007. Effects of sewage sludge amendment on heavy metal accumulation and consequent responses of *Beta vulgaris* plants. *Chemosphere.* 67: 2229-2240.

Skorzynska, E., Baszynski, T., 1993. The changes in PSII complex polypeptides under cadmium treatment – are they of direct or indirect nature? *Acta Physiologia Plantarum.* 15(4): 263–269.

Slebos, R.J.C., Li, M., Evjen, A.N., Coffa, J., Shyr, Y. Yarbrough, W.G., 2006. Mutagenic effect of cadmium on tetranucleotide repeats in human cells. *Mutation Research.* 602; 92-99.

Speir, T.W., Van Schaik, A.P., Percival, H.J., Close, M.E. Pang, L.P., 2003. Heavy metals in soil, plants and groundwater following high-rate sewage sludge application to land. *Water Air Soil Pollution.* 150: 319-358.

Steen Redeker, E., van Campenhout, K., Bervoets, L., Reijnders, H. Blust, R., 2007. Subcellular distribution of Cd in the aquatic oligochaete *Tubifex tubifex*, implications for trophic availability and toxicity. *Environmental Pollution.* 148: 166-175.

Steffens, J.C., 1990. The Heavy Metal-Binding Peptides of Plants. *Annual Review of Plant Physiology and Plant Molecular Biology.* 41: 553-575.

Steinkellner, H., Mun-Sik, K., Helma, C., Ecker, S., Ma, T.H., Horak, O., Kundi, M. Knasmüller, S., 1998. Genotoxic effects of heavy metals: Comparative investigation with plant bioassays. *Environmental and Molecular Mutagenesis.* 31: 183-191.

Sung-Hyun, K., Jin-Sung, P., In-Sook, L., 2009. Characterization of Cadmium-Binding Ligands from Roots of *Echinochloa crusgalli* var. frumentacea. *Journal of Plant Biology.* 52:167–170.

Sunkar, R., Bartels, D., Kirch, H.H., 2003. Overexpression of a stress-inducible aldehyde dehydrogenase gene from *Arabidopsis thaliana* in transgenic plants improves stress tolerance. *Plant Journal.* 35(4):452-64.

Suzuki, N., Yamaguchi, Y., Koizumi, N., Sano, H., 2002. Functional characterization of a heavy metal binding protein CdI19 from arabidopsis. *The Plant Journal.* 32(2):165-73.
Tamás, L., Dudíková, J., Durceková, K., Halusková, L., Huttová, J., Mistrík, I., 2009. Effect of cadmium and temperature on the lipoxygenase activity in barley root tip. *Protoplasma.* 235(1-4):17-25.
Tanhuanpää, P., Kalendar, R., Schulman, A.H., Kiviharju, E., 2007. A major gene for grain cadmium accumulation in oat (*Avena sativa* L.). *Genome.* 50(6):588-94.
Thomine, S., Wang, R., Ward, J.M., Crawford, N.M., Schroeder, J.I., 2000. Cadmium and iron transport by members of a plant metal transporter family in Arabidopsis with homology to Nramp genes. *Proceedings of the National Academy of Sciences.* 97(9):4991-6.
Tiryakioglu, M., Eker, S., Ozkutlu, F., Husted, S., Cakmak, I., 2006. Antioxidant defence system and cadmium uptake in barley genotypes differing in cadmium tolerance. *Journal of Trace Elements in Medicine and Biology.* 20(3):181-9.
Tommasini, R., Vogt, E., Fromenteau, M., Hörtensteiner, S., Matile, P., Amrhein, N., Martinoia, E., 1998. An ABC-transporter of *Arabidopsis thaliana* has both glutathione-conjugate and chlorophyll catabolite transport activity. *The Plant Journal.* 13(6):773-80.
Ueno, D., Ma, J.F., Iwashita, T., Zhao, F.J., McGrath, S.P., 2005. Identification of the form of Cd in the leaves of a superior Cd-accumulating ecotype of *Thlaspi caerulescens* using Cd-113-NMR. *Planta.* 221: 928-936.
Unyayar, S., Celik, A., Cekiç, F.O., Gözel, A., 2006. Cadmium-induced genotoxicity, cytotoxicity and lipid peroxidation in *Allium sativum* and *Vicia faba. Mutagenesis.* 21(1):77-81.
Vaipayee, P., Dhawan, A., Shanker, R., 2006. Evaluation of the alkaline comet assay conducted with the wetlands plant *Bacopa monnieri* L. as a model for ecogenotoxicity assessment. *Environmental and Molecular Mutagenetics.* 47: 483-489.
Valverde, M., Trejo, C., Rojas, E., 2001. Is the capacity of lead acetate and cadmium chloride to induce genotoxic damage due to direct DNA-metal interaction? *Mutagenesis.* 16: 265-270.
van de Mortel, J.E., Schat, H., Moerland, P.D., Ver Loren van Themaat, E., van der Ent, S., Blankestijn, H., Ghandilyan, A., Tsiatsiani, S., Aarts, M.G., 2008. Expression differences for genes involved in lignin, glutathione and sulphate metabolism in response to cadmium in

Arabidopsis thaliana and the related Zn/Cd-hyperaccumulator *Thlaspi caerulescens*. *Plant Cell and Environment.* 31(3):301-24.

van Rossum, F., Bonnin, I., Fenart, S., Pauwels, M., Petit, D., Saumitou-Laprade, P., 2004. Spatial genetic structure within a metallicolous population of *Arabidopsis halleri*, a clonal, self-incompatible and heavy-metal-tolerant species. *Molecular Ecology.* 13(10):2959-67.

Vassilev, A., Yordanov, I., 1997. Reductive analysis of factors limiting growth of cadmium-treated plants: a review. *Bulgarian Journal of Plant Physiology.* 23: 114-133.

Vassilev, A., Vangronsveld, J., Yordanov, I., 2002. Cadmium phytoextraction: present state, biological backgrounds and research needs. *Bulgarian Journal of Plant Physiology.* 28(3–4): 68–95.

Vijver, M.G., Koster, M., Peijnenburg, W.J.G.M., 2007. Impact of pH on Cu accumulation kinetics in earthworm cytosol. *Environmental Science Technology.* 41: 2255-2260.

Vijver, M.G., van Gestel, C.A.M., Lanno, R.P., van Straalen, N.M. Peijnenburg, W.J.G.M., 2004. Internal metal sequestration and its ecotoxicological relevance: a review. *Environmental Science Technology.* 38: 4705-4712.

Vijver, M.G., van Gestel, C.A.M., van Straalen, N.M., Lanno, R.P. Peijnenburg, W.J.G.M., 2006. Biological significance of metals partitioned to subcellular fractions within earthworms (*Aporrectodea caliginosa*). *Environmental Toxicological Chemistry.* 807-814.

Villatoro-Pulido, M., Font, R., De Haro-Bravo, M.I., Romero-Jiménez, M., Anter, J., De Haro Bailón, A., Alonso-Moraga, A., Del Río-Celestino, M., 2009. Modulation of genotoxicity and cytotoxicity by radish grown in metal-contaminated soils. *Mutagenesis.* 24(1):51-7.

Vogeli-Lange, R., Wagner, G.J., 1990. Subcellular localization of cadmium and cadmium-binding peptides in tobacco leaves: implication of a transport function for cadmium-binding peptides. *Plant Physiology.* 92: 1086-1093.

Waalkes, M.P., 2003. Cadmium carcinogenesis. *Mutation Research.* 533: 107-120.

Wagner, G.J., 1993. Accumulation of cadmium in crop plants and its consequences to human health. *Advances in Agronomy.* 51: 173-212.

Wallace, W.G., Lee, B.G., Luoma, S.N., 2003. Subcellular compartmentalization of Cd and Zn in two bivalves. I. Significance of metal-sensitive fractions (MSF) and biologically detoxified metal (BDM). *Marine Ecology Progress Series.* 249: 183-197.

Wallace, W.G., Lopez, G.R., 1997. Bioavailability of biologically sequestered cadmium and the implications of metal detoxification. *Marine Ecology Progress Series.* 147: 149-157.
Wallace, W.G., Lopez, G.R. Levinton, J.S., 1998. Cadmium resistance in an oligochaete and its effect on cadmium trophic transfer to an omnivorous shrimp. *Marine Ecology Progress Series.* 172: 225-237.
Wallace, W.G., Luoma, S.N., 2003. Subcellular compartmentalization of Cd and Zn in two bivalves. II. Significance of trophically available metal (TAM). *Marine Ecology Progress Series.* 257: 125-137.
Wang, W.-X., Fisher, N.S., 1999. Assimilation efficiencies of chemical contaminants in aquatic invertebrates: a synthesis. *Environmental Toxicology and Chemistry.* 18: 2034-2045.
Wang, W.X., Rainbow, P.S., 2006. Subcellular partitioning and the prediction of cadmium toxicity to aquatic organisms. *Environmental Chemistry.* 3: 395-399.
Watanabe, T., Misawa, S., Hiradate, S., Osaki, M., 2008. Root mucilage enhances aluminium accumulation in *Melastoma malabathricum*, an aluminium accumulator. *Plant Signal Behaviour.*3(8):603-5.
Xie, L., Funk, D.H., Buchwalter, D.B., 2009. Trophic transfer of Cd from natural periphyton to the grazing mayfly *Centroptilum triangulifer* in a life cycle test. *Environmental Pollution.*
Xu, J., Chai, T., Zhang, Y., Lang, M., Han, L.., 2009. The cation-efflux transporter BjCET2 mediates zinc and cadmium accumulation in *Brassica juncea* L. leaves. *Plant and Cell Report.* 28(8):1235-42.
Yu, R.Q., Fleeger, J.W., 2006. Effects of nutrient enrichment, depuration substrate, and body size on the trophic transfer of cadmium associated with microalgae to the benthic amphipod *Leptocheirus plumulosus*. *Environmental and Toxicological Chemistry.* 25(11):3065-72.
Zenk, M.H., 1996. Heavy metal detoxification in higher plants: a review. *Gene.* 179: 21-30.
Zhang, C., Qiu, B.S., 2007. Reactive oxygen species metabolism during the cadmium hyperaccumulation of a new hyperaccumulator *Sedum alfredii* (Crassulaceae). *Journal of Environmental Science. (China)* 19(11):1311-7.
Zhang, Y., Xiao, H., 1998. Antagonistic effect of calcium, zinc and selenium against cadmium induced chromosomal aberrations and micronuclei in root cells of *Hordeum vulgare*. *Mutation Research.* 420(1-3):1-6.
Zhang, Y., Yang, X., 1994. The toxic effects of cadmium on cell division and chromosomal morphology of *Hordeum vulgare*. *Mutation Research.* 312(2):121-6.

Zhang, X.H., Lin, A.J., Su, Y.H., Zhu, Y.G., 2006. DNA damages and apoptosis induced by cd in the leaves of horsebean *Vicia faba*. *Huan Jing Ke Xue.* 27(4):787-93.

Zhao, F.J., Lombi, E., McGrath, S.P., 2003. Assessing the potential for zinc and cadmium phytoremediation with the hyperaccumulator *Thlaspi caerulescens*. *Plant. Soil.* 249: 37-43.

Zhu, R., Macfie, S.M., Ding, Z., 2005 Cadmium-induced plant stress investigated by scanning electrochemical microscopy. *Journal of Experimental Botany.* 56 (421), 2831–2838.

Zhuang, P., Zou, H., Shu, W., 2009. Biotransfer of heavy metals along a soil-plant-insect-chicken food chain: field study. *Journal of Environmental Science. (China)* 21(6):849-53.

Zienolddiny, S., Svendsrud, D.H., Ryberg, D., Mikalsen, A.B., Haugen, A., 2000. Nickel (II) induces microsatellite mutations in human lung cancer cell lines. *Mutation Research.* 452: 91-100.

INDEX

A

acid, 13, 16, 17, 25, 31, 32, 44, 51, 52
adenine, 19
adenosine, 4
age, 8, 23, 38
agriculture, 39
aldehydes, 15
aldolase, 11
alfalfa, 14, 43, 55
aluminum, 53
amino acids, 5, 31
aneuploidy, 18
anhydrase, 1, 11, 30
anthocyanin, 48
antioxidant, 8, 14, 38, 42, 44, 46, 51, 57
apoptosis, 24, 38, 63
applications, 36
Arabidopsis thaliana, 14, 21, 43, 46, 59, 60, 61
ARC, 3, 18, 33
ascorbic acid, 16
ash, 27, 46, 50, 53, 56
assessment, 9, 12, 24, 27, 41, 44, 47, 52, 54, 60
assimilation, 12, 35, 36, 37, 43, 58
ATP, 15
authors, 9, 15, 16, 18, 19, 23, 24, 26, 27
availability, 4, 15, 35, 48, 59

B

backcross, 23
background, 27, 48
barley, 10, 14, 21, 27, 52, 60
barriers, 34, 54
binding, 4, 13, 15, 19, 31, 35, 36, 60, 61
bioaccumulation, 1, 33, 34, 39, 50, 54
bioassay, 22, 48
bioavailability, 3, 34, 37, 53, 54
biochemistry, 51, 55
bioindicators, 44
biomarkers, 55
biomass, 7, 17, 43
biomonitoring, 27
biosynthesis, 9, 14, 15, 42, 55, 57
biosynthetic pathways, 9
biotic, 12
body size, 62
burning, 1

C

cabbage, 48
cadmium, vii, 1, 18, 37, 38, 41, 42, 43, 44, 45, 46, 47, 48, 49, 51, 52, 53, 54, 55, 56, 57, 58, 59, 60, 61, 62, 63
calcium, 4, 45, 57, 62
cancer, 18
candidates, 17

carbohydrates, vii, 11
carbon, vii, 8, 37, 58, 59
carcinogenesis, 61
carrier, 16, 51
case study, 53, 54
cation, 16, 19, 31, 32, 33, 58, 62
cDNA, 16, 17, 23, 44, 45, 47
cell, vii, 4, 5, 9, 11, 12, 13, 14, 16, 17, 18, 22, 25, 26, 27, 31, 38, 43, 47, 48, 49, 62, 63
cell culture, 49
cell cycle, 26, 27, 38
cell death, 9, 13, 16, 43
cell line, 18, 22, 63
cell lines, 18, 22, 63
cell membranes, 11
cellular homeostasis, 31
centromere, 25
chain transfer, 35
channels, 5, 17
chicken, 34, 63
chlorophyll, vii, 9, 11, 42, 49, 53, 60
chloroplast, 11, 12, 16, 22, 37, 48, 54
chromium, 45
chromosome, 18, 22, 25, 26, 53, 54, 56
circulation, 31
classes, 17, 46
cloning, 16, 51
coal, 27, 51, 53
coding, 38
coefficient of variation, 27
coffee, 48
combustion, 51
competition, 4
compilation, 19, 50
complement, 14
complexity, 34
components, 14, 17
composition, 15, 31, 54
compounds, 7, 14, 16, 18, 24
concentration, 3, 17, 19, 26, 27, 29, 36, 37, 38
conditioning, 4, 37
Congress, 48, 54
conservation, 7

consumers, 33, 34, 36
consumption, 34
contaminated soils, 3, 23, 33, 50, 61
contamination, 1, 7, 37, 50
control, 14, 16, 20, 27, 34, 38
conversion, 14
coordination, 38
copper, 1, 44, 46, 55, 57
correlation, 26
cortex, 4
crops, vii, 3, 7, 56
cycling, 34
cytokinesis, 25
cytometry, viii, 20, 26, 27, 38, 54
cytoplasm, 16, 25
cytoskeleton, 17
cytotoxicity, 2, 30, 47, 60, 61

D

damages, 18, 45
death, 7
defence, 13, 15, 38, 42, 60
deficiency, 9, 10, 12, 51, 58
degradation, vii, 9, 11, 17, 27, 44
density, 11
Department of the Interior, 50
dephosphorylation, 17
deposits, 35
destruction, 11
detection, 18, 19, 22, 48, 55
detoxification, vii, 13, 14, 16, 17, 31, 32, 34, 35, 39, 45, 49, 51, 62
differentiation, 22, 38
diffusion, 4, 31
disorder, 11
dispersion, 27
displacement, 13
dissociation, 10
distribution, vii, 30, 35, 36, 44, 49, 54, 56, 59
diversity, 53
division, 12, 25, 62

DNA, vii, 17, 18, 21, 22, 23, 24, 25, 26, 27, 35, 38, 42, 43, 46, 47, 49, 50, 52, 55, 57, 60, 63
DNA damage, 22, 24, 25, 42, 47, 50, 52, 63
DNA repair, 18, 43
DNA strand breaks, 50
draft, 41
duration, 8, 34

E

earthworm, 61
electron, 8, 10, 12, 13, 16, 32, 37, 49, 58
encoding, 15, 23
endoderm, 4
energy, 8
engineering, 17, 45
environment, vii, 1, 8, 29, 33, 55
environmental conditions, 36
Environmental Protection Agency, vii, 1
environmental quality, 7
enzymes, vii, 8, 11, 13, 14, 15, 16, 17, 35, 37, 38, 42, 57, 58
epidermis, 4
ERA, 38
EST, 23
ethylene, 15
European Commission, 41
evolution, 10
excitation, 12, 26
experimental condition, 38
exposure, 12, 14, 15, 16, 18, 22, 23, 25, 27, 31, 33, 34, 36, 39, 43, 50, 51, 55
expressed sequence tag, 23

F

family, 31, 47, 60
fertilizers, 1, 4, 49
floating, 41, 57
fluorescence, vii, 9, 26, 42, 49, 51, 53
focusing, 22
food, vii, 1, 30, 33, 34, 35, 39, 45, 53, 54, 63

fossil, 1
fragments, 21, 23, 25, 46
freshwater, 45, 55, 56
fructose, 11

G

gene, 13, 15, 17, 18, 21, 23, 38, 42, 47, 48, 50, 52, 53, 56, 58, 59, 60
gene expression, 13, 15, 17, 38, 42, 58
generation, 25
genetic linkage, 22
genetic marker, 19, 38
genetics, 22
genome, 18, 38, 48
germination, 23
glutathione, 14, 15, 42, 53, 60
god, 50
grana, 11
granules, 32, 35, 36
grass, 51
grazing, 62
groundwater, 59
groups, 11, 13
growth, vii, 7, 8, 11, 12, 15, 16, 29, 37, 42, 44, 46, 49, 51, 52, 55, 56, 57, 58, 61
guanine, 19

H

half-life, 1
haploid, 18
health, 3, 26, 30, 41, 43, 61
health effects, 43
heat, 35, 36
heavy metals, 39, 45, 46, 47, 48, 50, 59, 63
heterogeneity, 24
histidine, 31, 58
histogram, 26
homeostasis, 4, 16, 31, 39, 45, 55
human exposure, 33
hydrophyte, 41
hydroponics, 52
hypersensitivity, 15

I

ideal, 19
imbalances, 42
in situ hybridization, 25
in vivo, 27, 55, 57
indices, 52
indirect effect, 8
inducer, 31
induction, 18, 25, 43, 44, 57
industry, 1, 49
ingestion, 34
inhibition, 7, 10, 11, 12, 16, 19, 49, 51, 56
injury, iv, 38
instability, 19, 22, 54, 55
integration, 18
interaction, 16, 37, 49, 59, 60
intervention, 18
invertebrates, 33, 53, 56, 62
ion channels, 5
ion transport, 42
ions, 4, 5, 9, 10, 13, 16, 31, 38, 49
iron, 1, 12, 58, 59, 60
iron transport, 60
isopods, 33, 36
isozymes, 16

K

kidney, 21
kinetics, 44, 61

L

labeling, 25
land, 59
lesions, 8, 37
life cycle, 62
ligand, 30, 31, 39
lignin, 15, 60
lipid peroxidation, 15, 44, 60
lipids, 11, 13, 55
localization, 13, 61
locus, 23

lung cancer, 22, 63
lymphocytes, 42

M

magnesium, 12
maintenance, 15
majority, 31
management, 34, 44
manganese, 12, 55
manufacturing, 49
mapping, 23, 52
measurement, 27, 55
measures, 24
membranes, 4, 11, 15, 16
menadione, 57
mercury, 1, 49, 55
meristem, 25, 46, 47, 56
mesophyll, 10, 11, 32
metabolic pathways, 8, 16
metabolism, vii, 7, 8, 9, 11, 15, 16, 17, 37, 41, 48, 53, 57, 58, 59, 60, 62
metal salts, 46
metals, vii, 1, 4, 5, 7, 9, 11, 12, 13, 18, 19, 20, 21, 22, 25, 27, 29, 30, 31, 33, 34, 35, 36, 39, 45, 47, 50, 53, 56, 59, 61
methodology, 27
mice, 49
micronucleus, 24, 25, 38, 48, 57
micronutrients, 1, 4, 11, 31
microsatellites, vii, 38, 54
microscopy, 63
migration, 24
milligrams, 27
mining, 1, 3
mitochondria, 13, 16, 43
mitosis, 25
mitotic index, 19
mobility, 1, 3, 4
model, vii, 16, 18, 19, 24, 30, 34, 36, 38, 39, 41, 46, 60
model system, 39
molecular weight, 31, 35
molecules, 9, 13, 14, 38, 48
morphology, 7, 62

mortality, 56
mortality rate, 56
mother cell, 20
movement, 4, 5
mutagen, 18, 50
mutagenesis, 13, 18
mutant, 15, 18, 49
mutation, 50
mutation rate, 50

N

network, 14, 50
nickel, 22, 55, 58
nitric oxide, 57
nitrogen, 7, 44
NMR, 60
nucleic acid, 13, 19
nucleotides, 49
nucleus, 17, 25, 47, 52
nutrients, 4, 8, 11, 16, 37, 51

O

OECD, 12, 55
oilseed, 11, 42, 44
organ, 20, 21
organelles, 16, 35
organic matter, 1, 3, 37
organism, 26, 35, 36
oxidation, 13, 16
oxidative damage, 13
oxidative stress, vii, 7, 8, 13, 15, 16, 24, 25, 38, 41, 49, 57, 58
oxygen, 10, 12, 13, 38, 41, 62
ozone, 46

P

parameter, 12, 39
particles, 26
partition, 34
pathways, 17, 29, 34, 37, 38, 39
peptides, 31, 61
peripheral blood, 42
permeability, 42, 45
petroleum, 43
phenotype, 18
phosphorus, 49
phosphorylation, 13, 17
photosynthesis, vii, 7, 8, 9, 10, 11, 12, 16, 42, 46, 51, 52, 58
phytoplankton, 34
phytoremediation, 17, 29, 30, 39, 53, 63
plasma, 4, 16, 49
plasma membrane, 4, 16, 49
plasticity, 50
ploidy, 26, 47, 54
point mutation, vii, 19, 22, 23
pollen, 20
pollutants, 1, 37
pollution, vii, 3, 8, 19, 21, 39, 46, 47, 52
polonium, 53
polycyclic aromatic hydrocarbon, 27
polymorphism, 19, 21, 24
pools, 36
poor, 7
population, 22, 24, 26, 61
potassium, 12
prediction, 62
pressure, 22
prevention, 18
probe, 25
producers, 33
production, 7, 13, 15, 16, 19, 38, 49
productivity, 7
proliferation, 38, 47
properties, 30
prostate, 18
protective role, 47
proteins, 11, 13, 16, 17, 31, 35, 36, 37, 38
proteomics, 37

Q

quantum yields, 9

R

radicals, 46
rain, 4, 25, 52
range, 1, 51
rape, 11, 42, 44, 47
reactions, 10, 12, 13, 16
reactive oxygen, 12, 41, 43, 46, 47, 49, 57
recombination, 19, 21, 47
redistribution, 30
region, 25, 56
regulation, 17, 37, 57
relationship, 11, 24, 35, 52
relevance, vii, 61
repair, 17, 18, 38, 49, 50
replication, 22, 38, 48
residues, 31, 51
resistance, 15, 23, 62
resources, 7
respiration, 7, 12
rice, 3, 21, 27, 49, 52
risk, 19, 33, 39
risk assessment, 19

S

safety, 7
salinity, 33
scatter, 26
scavengers, 13, 14
secretion, 23, 52
seed, 46
seedlings, 20, 21, 43, 44, 47, 49, 52, 53, 55, 57
selenium, 16, 47, 62
senescence, 7, 9, 15
sensitivity, 10, 24
sequestration proteins, 35
sewage, 1, 49, 59
shape, 11
shoot, 12, 16, 48
shrimp, 35, 51, 62
signal transduction, 17, 41
signalling, 13
signals, 38
sludge, 1, 3, 42, 49, 56, 59
sodium, 4, 43
soil, 3, 4, 7, 27, 29, 34, 37, 41, 42, 44, 48, 49, 52, 53, 59, 63
solid waste, 20, 42, 46, 56
solubility, 53
somatic mutations, 47
soybean, 44
species, vii, 7, 9, 10, 12, 13, 14, 15, 19, 22, 24, 25, 29, 30, 33, 34, 35, 38, 39, 41, 43, 46, 47, 49, 50, 57, 61, 62
speed, 17
stability, 54
stele, 4, 5
stimulus, 38
storage, 32, 34, 36, 39, 44
strategy, 16, 44, 46
stress, vii, 8, 9, 12, 13, 14, 15, 17, 24, 25, 38, 42, 43, 46, 47, 48, 52, 53, 54, 55, 56, 57, 58, 59, 63
stressors, 22
stroma, 11
strong interaction, 9
substitution, 9
sugar, 9, 51
sugar beet, 9, 51
supply, 14
surfactant, 25, 52
survival, 38, 51
suspensions, 26
symptoms, 8, 9, 30, 37
synthesis, 8, 9, 37, 46, 62

T

targets, 9, 17
telophase, 21, 56
temperature, 3, 33, 37, 60
testing, 27, 50, 56
threat, 33
threshold, 29
thymine, 19
tissue, 4, 8, 26, 34
tobacco, 24, 32, 47, 61

toxic effect, 18, 62
toxic metals, 11
toxic substances, 41
toxicity, vii, 1, 3, 4, 7, 8, 9, 10, 12, 13, 17, 25, 34, 35, 36, 38, 41, 43, 44, 46, 49, 51, 52, 55, 56, 57, 59, 62
toxicology, 18, 21, 30, 57
toxicology studies, 18
traits, 19, 23, 46
transcription, 17, 37
transcripts, 18, 23, 37
transduction, 17
transformation, 9
translocation, 12, 29
transpiration, 5, 11
transport, 4, 5, 8, 11, 15, 17, 23, 29, 30, 31, 37, 39, 45, 48, 58, 60, 61
tumours, 18

U

ultrastructure, 11

V

vacuole, 31, 32
variability, 22, 30
variations, 30, 55

vascular system, 4
vegetables, 3
visualization, 57

W

wastewater, 43, 56
water policy, 41
wetlands, 60
wheat, 11, 21, 22, 46, 53, 54, 56
workers, 33
worms, 35

X

xylem, 4, 5

Y

yeast, 15, 45

Z

zinc, 4, 44, 46, 48, 50, 62, 63
zooplankton, 34